CLIMATE CHANGE AND ITS CAUSES, EFFECTS AND PREDICTION

CLIMATE CHANGE

SELECT FEDERAL EFFORTS TO MONITOR IMPACTS ON WATER INFRASTRUCTURE AND RESOURCES

Climate Change and its Causes, Effects and Prediction

Additional books in this series can be found on Nova's website under the Series tab.

Additional e-books in this series can be found on Nova's website under the e-book tab.

CLIMATE CHANGE AND ITS CAUSES, EFFECTS AND PREDICTION

CLIMATE CHANGE

SELECT FEDERAL EFFORTS TO MONITOR IMPACTS ON WATER INFRASTRUCTURE AND RESOURCES

AILEEN STABLUM
EDITOR

nova publishers
New York

For permission to use material from this book please contact us:
Telephone 631-231-7269; Fax 631-231-8175
Web Site: http://www.novapublishers.com

Library of Congress Cataloging-in-Publication Data

ISBN: 978-1-63117-584-8

Published by Nova Science Publishers, Inc. † New York

CONTENTS

PREFACE

Climate change is having a variety of impacts on natural resources in the United States, ranging from more severe drought to increased flooding, and is altering assumptions that have been central to water resource planning and management. The Department of Defense's U.S. Army Corps of Engineers (Corps) and the Department of the Interior's Bureau of Reclamation (Reclamation) have assessed water resource and infrastructure vulnerabilities and taken steps to develop guidance and strategies to adapt to the effects of climate change. This book examines actions taken by the Corps and Reclamation since 2009 to assess and respond to the potential effects of climate change on water infrastructure; and challenges, if any, faced by the Corps and Reclamation in assessing and responding to the potential effects of climate change on water infrastructure, and the steps the agencies are taking to address them.

Chapter 1 – The effects of climate change on water resources have already been observed and are expected to continue. The Corps and Reclamation own and operate key water resource management infrastructure, such as dams and reservoirs. Adaptation— adjustments in natural or human systems to a new or changing environment that exploits beneficial opportunities or moderates negative effects—can be used to help manage the risks to vulnerable resources. In 2009, a law—commonly referred to as the SECURE Water Act—and a presidential executive order directed federal agencies to address the potential impacts of climate change.

GAO was asked to review agency actions to address climate change impacts on water infrastructure. This report examines (1) actions taken by the Corps and Reclamation since 2009 to assess and respond to the potential effects of climate change on water infrastructure and (2) challenges, if any, faced by the Corps and Reclamation in assessing and responding to the potential effects of climate change on water infrastructure, and the steps the agencies are taking to address them. GAO analyzed the agencies' climate change adaptation guidance and planning documents and interviewed agency officials and other key stakeholders, including water users, environmental groups, and researchers.

GAO is not making any recommendations.

Chapter 2 – Of all the potential threats posed by climatic variability and change, those associated with water resources are arguably the most consequential for both society and the environment (Waggoner, 1990). Climatic effects on agriculture, aquatic ecosystems, energy, and industry are strongly influenced by climatic effects on water. Thus, understanding changes in the distribution, quantity and quality of, and demand for water in response to climate variability and change is essential to planning for and adapting to future climatic conditions. A central role of the U.S. Geological Survey (USGS) with respect to climate is to document environmental changes currently underway and to develop improved capabilities to predict future changes. Indeed, a centerpiece of the USGS role is a new Climate Effects Network of monitoring sites. Measuring the climatic effects on water is an essential component of such a network (along with corresponding effects on terrestrial ecosystems).

The USGS needs to be unambiguous in communicating with its customers and stakeholders, and with officials at the Department of the Interior, that although modeling future impacts of climate change is important, there is no more critical role for the USGS in climate change science than that of measuring and describing the changes that are currently underway. One of the best statements of that mission comes from a short paper by Ralph Keeling (2008) that describes the inspiration and the challenges faced by David Keeling in operating the all-important Mauna Loa Observatory over a period of more than four decades. Ralph Keeling stated: "The only way to figure out what is happening to our planet is to measure it, and this means tracking changes decade after decade and poring over the records."

Chapter 3 – This report responds to the requirements of Section 9506 of the Omnibus Public Lands Act [Public Law (PL) 111–11; Appendix A] calling for a report to Congress that describes the impacts of global climate change on freshwater resources of the United States, and identifies key actions to improve the Nation's capacity to detect and predict changes in freshwater resources that are likely to result from a changing climate.

The steps described in the report are intended to help decision-makers and water resource managers by facilitating improvements in observational data, data acquisition, and modeling capabilities.

In: Climate Change
Editor: Aileen Stablum

ISBN: 978-1-63117-584-8

Chapter 1

CLIMATE CHANGE: FEDERAL EFFORTS UNDER WAY TO ASSESS WATER INFRASTRUCTURE VULNERABILITIES AND ADDRESS ADAPTATION CHALLENGES*

United States Government Accountability Office

WHY GAO DID THIS STUDY

The effects of climate change on water resources have already been observed and are expected to continue. The Corps and Reclamation own and operate key water resource management infrastructure, such as dams and reservoirs. Adaptation— adjustments in natural or human systems to a new or changing environment that exploits beneficial opportunities or moderates negative effects—can be used to help manage the risks to vulnerable resources. In 2009, a law—commonly referred to as the SECURE Water Act—and a presidential executive order directed federal agencies to address the potential impacts of climate change.

GAO was asked to review agency actions to address climate change impacts on water infrastructure. This report examines (1) actions taken by the Corps and Reclamation since 2009 to assess and respond to the potential

* This is an edited, reformatted and augmented version of the United States Government Accountability Office publication, GAO-14-23, dated November 2013.

effects of climate change on water infrastructure and (2) challenges, if any, faced by the Corps and Reclamation in assessing and responding to the potential effects of climate change on water infrastructure, and the steps the agencies are taking to address them. GAO analyzed the agencies' climate change adaptation guidance and planning documents and interviewed agency officials and other key stakeholders, including water users, environmental groups, and researchers.

GAO is not making any recommendations.

What GAO Found

The Department of Defense's U.S. Army Corps of Engineers (Corps) and the Department of the Interior's Bureau of Reclamation (Reclamation) have assessed water resource and infrastructure vulnerabilities and taken steps to develop guidance and strategies to adapt to the effects of climate change. Specifically, since 2009, the Corps has completed a high-level assessment of the vulnerabilities to climate change of various agency missions. The assessment found, for example, that the effects of increasing air temperatures on glaciers could negatively impact mission areas including navigation and flood damage reduction. The Corps has also conducted pilot studies to help identify adaptation guidance and strategies; it has completed 5 of the 15 pilot studies initiated and plans to start another study in 2013. Similarly, Reclamation has completed baseline assessments of the potential impacts of climate change on future water supplies for the major river basins where it owns and operates water management infrastructure. Reclamation, in collaboration with nonfederal entities, is now conducting more focused assessments, known as Basin Studies, through which Reclamation seeks to identify water supply vulnerabilities and project future climate change impacts on the performance of water infrastructure. According to agency officials, these studies will also help Reclamation develop adaptation strategies to address these impacts, such as operational or physical changes to existing water infrastructure or development of new facilities. Three Basin Studies have been completed, an additional 14 are under way, and 2 more are planned. Reclamation next plans to initiate feasibility studies for adaptation strategies identified in completed Basin Studies. Both agencies are incorporating what they have learned from their efforts into agency policies, planning, and guidance, according to agency officials.

In 2009, the Corps, Reclamation, the National Oceanic and Atmospheric Administration, and the U.S. Geological Survey (USGS), jointly published a study that identified several challenges that climate change poses for water resource managers, and the Corp and Reclamation are collaboratively addressing these challenges. Specifically, these agencies are

- identifying the data and tools needed by water managers to address climate change, which will help guide federal research efforts;
- obtaining needed climate data by collaborating with other agencies to help ensure that the data are collected, such as by sharing some costs associated with maintaining USGS's stream flow measurement activities, which are valuable to Corps water planning and management;
- integrating climate science into water resource management decision making through activities such as developing and communicating science to inform climate adaptation strategies; and
- collaborating in the development of a climate change science training program for federal and nonfederal water resources managers.

The Corps and Reclamation have collaborated together and with others in a manner that is generally consistent with practices that GAO has identified as important to enhancing and sustaining collaboration among agencies. The Corps and Reclamation have made collaboration a key element of their policy and plans for adapting to the effects of climate change and have reinforced accountability for collaboration through agency performance management systems.

ABBREVIATIONS

CCAWWG	Climate Change and Water Working Group
Corps	U.S. Army Corps of Engineers
Interior	U.S. Department of the Interior
NOAA	National Oceanic and Atmospheric Administration
Reclamation	Bureau of Reclamation
Task Force	Interagency Climate Change Adaptation Task Force
USGS	U.S. Geological Survey
WWCRA	West-Wide Climate Risk Assessment

November 14, 2013

The Honorable Barbara Boxer
Chairman
Committee on Environment and Public Works
United States Senate

The Honorable Max Baucus
United States Senate

The Honorable Dick Durbin
United States Senate

The Honorable Harry Reid
United States Senate

The Honorable Mark Udall
United States Senate

The Honorable Tom Udall
United States Senate

According to recent scientific studies, including those conducted by the U.S. Global Change Research Program,[1] climate-related changes— including increases in air and water temperatures, changes in precipitation patterns, reduced snow cover, retreating glaciers, and rising sea levels—will likely adversely affect many aspects of the natural environment in the United States.[2] The U.S. Department of the Interior's (Interior) U.S. Geological Survey (USGS) has reported that of all the potential threats posed by climate change, those associated with water resources are arguably the most consequential for both society and the environment.[3] The effects of climate change on water resources have already been observed and are expected to continue. For example, a shift in the form of precipitation from snow to rain and earlier melting of mountain snowpack has been documented in western states. This change in precipitation is expected to result in a decrease in the amount of reliable water supply in areas where snow has been a major source. As we reported in October 2009,[4] policymakers are increasingly viewing adaptation—defined by the National Research Council as adjustments in natural or human systems to a new or changing environment that exploits beneficial opportunities or moderates negative effects—as a risk- management strategy to protect vulnerable resources that might be affected by changes in the climate.[5] GAO added limiting the federal government's fiscal exposure by better managing climate change risks to our 2013 update of high-risk areas because of the government's role, among other things, as the owner or operator of extensive infrastructure.[6]

The Department of Defense's U.S. Army Corps of Engineers (Corps) and Interior's Bureau of Reclamation (Reclamation) own and operate key water resource management infrastructure, such as canals, dams, and reservoirs, that may be affected by climate change. The Corps is the largest and oldest federal water management agency, with water resources and infrastructure in every state authorized for specific purposes such as navigation, flood and coastal storm damage reduction, hydropower, and water supply, among other things. Established in 1902, Reclamation constructed dams, reservoirs, power plants, and canals in 17 western states, and today is the second largest producer of hydroelectric power in the western United States and the largest wholesaler of water in the nation. The Corps and Reclamation work with each other and with other federal agencies, such as the Department of Commerce's National Oceanic and Atmospheric Administration (NOAA) and USGS, which collect and interpret weather and climate information that is valuable for effective water resource management. In addition, the Corps and Reclamation coordinate with state and local water resource managers and other stakeholders who are responsible for water resources activities within their jurisdictions to manage water for agricultural, municipal, and conservation uses, among others.

Congress and the Executive Office of the President have directed federal agencies to address the potential impacts of climate change. In 2009, a law—commonly referred to as the SECURE Water Act—was enacted, requiring, among other things, that Reclamation establish a climate change adaptation program to (1) assess the effect of and risk resulting from global climate change on the quantity of water resources and (2) develop strategies to address potential water shortages and other impacts.[7] Reclamation was also required to report to relevant congressional committees its progress implementing the act within 2 years of passage and every 5 years thereafter. Also in 2009, the President signed Executive Order 13514,[8] calling for federal agencies to participate in the existing Interagency Climate Change Adaptation Task Force (Task Force). The executive order directed the Task Force to develop federal recommendations for adapting to climate change impacts both domestically and internationally and to recommend key components to include in a national strategy. Based on the Task Force recommendations, the President's Council on Environmental Quality issued implementing instructions directing federal agencies to, among other things, establish an agency climate change adaptation policy and deliver to the council and the Office of Management and Budget a climate adaptation plan for implementation in fiscal year 2013.[9]

You requested that we review the climate adaptation efforts of the Corps and Reclamation. This report examines (1) the actions the Corps and Reclamation have taken since 2009 to assess and respond to the potential effects of climate change on water infrastructure and (2) what challenges, if any, the Corps and Reclamation face in assessing and responding to the potential effects of climate change on water infrastructure, and the steps the agencies are taking to address them.

To describe what actions the Corps and Reclamation have taken to assess and respond to the potential effects of climate change since 2009, we reviewed and analyzed federal law, an executive order, and relevant guidance to identify the climate change impact assessment and adaptation activities that the Corps and Reclamation are directed to perform. We reviewed and analyzed the agencies' orders, policy, and program guidance and interviewed agency officials to determine how the Corps and Reclamation have interpreted and implemented federal laws and orders. We also reviewed and analyzed documentation, such as reports, studies, and plans, and interviewed agency officials, as well as other water resource managers and climate science specialists to gather information on what the Corps and Reclamation have done to assess and respond to the potential effects of climate change on existing and planned federal water infrastructure investments. We identified and selected the resource managers and climate science specialists using an iterative process, soliciting names from agency officials and others we interviewed, and selecting those identified as having a high level of participation and knowledge. From among those identified, we conducted semistructured interviews to obtain their views on the Corps' and Reclamation's efforts to assess and respond to the potential effects of climate change. We did not assess the Corps' and Reclamation's actions to implement specific climate change adaptation planning frameworks or steps because their efforts are in the early stages. We used in-house scientific expertise to analyze the soundness of the methodological approaches utilized in selected documents, and we determined them to be sufficiently sound for our purposes. Relevant assessments, reports, and studies are cited throughout this document.

To describe what challenges, if any, the Corps and Reclamation face in assessing and responding to the potential effects of climate change on water infrastructure, as well as steps these agencies are taking to address these challenges, we reviewed and analyzed agency documents, including planning documents and assessments of efforts to date. We also reviewed and analyzed documentation of actions taken by the Corps, Reclamation, and other water managers to help mitigate these challenges, and we assessed the extent to which these efforts have been collaborative based on prior GAO work.[10] In addition, we spoke with agency officials, as well as sought the perspectives of state and local water managers and climate specialists, to learn about the challenges faced by the Corps and Reclamation and the steps the agencies have taken to address them. We selected these managers and specialists using an iterative process, based on recommendations of agency officials and others we interviewed. From among those identified, we conducted semistructured interviews about challenges and potential steps to address them and conducted a content analysis of their responses.

We conducted this performance audit from October 2012 to November 2013 in accordance with generally accepted government auditing standards. Those standards require that we plan and perform the audit to obtain sufficient, appropriate evidence to provide a reasonable basis for our findings and conclusions based on our audit objectives. We believe that the evidence obtained provides a reasonable basis for our findings and conclusions based on our audit objectives.

BACKGROUND

Climate change is having a variety of impacts on natural resources in the United States, ranging from more severe drought to increased flooding, and is altering assumptions that have been central to water resource planning and management. Congress and the Executive Office of the President have directed federal agencies to address the potential impacts of climate change. The two key federal water resource management agencies included in this review—the Corps and Reclamation—have similar yet distinct roles in managing water for a wide variety of purposes. In addition to the water management challenges posed by climate change, both agencies are dealing with aging water management infrastructure with limited funding for maintenance and construction.

Potential Climate Change Impacts on Water Resources Can Be Addressed through Adaptation Planning

According to the U.S. Global Change Research Program, changes in the climate in the United States and its coastal waters have altered—and will continue to alter—the water cycle, affecting where, when, and how much water is available for all uses. Changes in the climate—including warmer temperatures, changes in precipitation patterns, rising sea levels, and more frequent and intense storms—affect water resources in a number of ways such as increased flooding in some areas and drought in others, and inundation and erosion in coastal areas. Precisely how and to what extent changes in the climate will affect particular water resources in the future is uncertain, but climate-related changes are expected to continue and increase in intensity in some areas of the nation.

Climate change has the potential to affect many aspects of the environment and society in which water resource management plays an active role. A 2011 federal interagency review of the potential impacts of climate change on water resources identified four interrelated areas of concern for water resource managers as follows[11]:

- assuring an adequate water supply for multiple needs, such as drinking water, agriculture, energy production, industrial uses, navigation, and recreation;
- protecting life, health, and property in the face of risks posed by climate change;
- protecting the quality of freshwater resources, including the quality of surface water and groundwater, and the health of fisheries and aquatic habitat; and
- protecting coastal and ocean resources as rising sea levels and changes in storm frequency, intensity, and duration impact coastal infrastructure.

Adaptation—defined by the National Research Council as adjustments in natural or human systems to a new or changing environment that exploits beneficial opportunities or moderates negative effects—is an element of the proposed responses to climate change that is gaining more attention. More specifically, policymakers are increasingly viewing adaptation as a risk-management strategy to protect vulnerable sectors and communities that might be affected. As we reported in our May 2013 report on land resource management agency adaptation efforts, climate change adaptation planning frameworks generally consist of four key elements that are reviewed and revised as needed as new information emerges.[12] These four elements are the following:

- Establish a mandate to address climate change with clearly articulated adaptation goals, objectives, and measures of success toward meeting goals.
- Assess and understand the risks, vulnerabilities, and opportunities posed by climate change by determining (1) what aspects of the climate are changing and over what periods, (2) which resources will be most at risk, (3) why these resources are likely to be vulnerable, and (4) what uncertainties are associated with the predicted climate change impacts and how this may impact adaptation efforts.
- Develop and prioritize management adaptation actions; that is, determine how to respond to the identified risks by considering a wide array of possible adaptation measures and identifying the highest priority adaptation measures.
- Implement management options, evaluate the results to determine the actions' effectiveness, and makes adjustments as necessary. As climate continues to change, adaptation actions need to be regularly monitored for effectiveness and plans need to incorporate new information about risks, lessons learned, and modified priorities.

According to a 2009 study collaboratively produced by the Corps, NOAA, Reclamation, and USGS,[13] a variety of water-management options might be considered to facilitate adaptation to climate change, including operational changes, demand management, and infrastructure changes. The study concluded that the options for responding to the effects of climate change will vary by location, and that evaluating options will likely require a partnership between federal, state, and local interests to attain consensus among water managers and users.

Federal Agencies Have Been Directed to Address Climate Change Adaptation

In 2009, Congress passed the SECURE Water Act, requiring Reclamation to establish a climate change adaptation program to (1) assess the effect of and risk resulting from global climate change on the quantity of water resources and (2) develop strategies to address potential water shortages and other impacts.[14] To assess the effect of climate change, the law requires Reclamation to analyze the extent to which changes in water supply will impact several areas, including Reclamation's ability to deliver water to its customers, hydroelectric power generation, fish and wildlife habitat, and recreation. The law requires Reclamation to assess specific risks to the water supply of each of its major river basins, including risks related to, among other things, changes in snow cover and the timing and quantity of runoff. To develop strategies to address impacts, the law requires Reclamation, in consultation with nonfederal stakeholders such as appropriate state water resource agencies, to consider and develop appropriate strategies to mitigate the impacts of water supply changes, such as strategies for modifying reservoir storage or operating guidelines and water conservation.[15]

Also in 2009, the President signed Executive Order 13514, Federal Leadership in Environmental, Energy, and Economic Performance which, among other things, directs agencies to participate in the Interagency Climate Change Adaptation Task Force (Task Force), which was already developing a strategy for adaptation to climate change, and to develop approaches through which the policies and practices of the agencies can be made compatible with the Task Force's strategy.[16] In October 2010, the Task Force delivered a progress report to the President through the Council on Environmental Quality containing overarching policy goals to advance climate adaptation and recommending the development of a national action plan to strengthen climate change adaptation for freshwater resources.[17] Based on the work of the Task Force, the Council on Environmental Quality subsequently issued detailed adaptation planning implementation instructions in March 2011.[18] The instructions directed the agencies to issue an agency-wide climate change adaptation policy statement, complete a high-level analysis of agency vulnerability to climate change, and submit to the Council on Environmental Quality and the Office of Management and Budget their climate adaptation plans by June 4, 2012, for implementation in fiscal year 2013.[19]

The Corps and Reclamation Construct, Operate, and Manage Federal Water Infrastructure

For over 230 years, the Corps has led the development and stewardship of much of the nation's public water resources. The Corps' Civil Works Program plans and manages water for transportation, recreation, energy, wildlife habitat, aquatic ecosystems, and water supply, while reducing the impacts of flood damages and other natural disasters. Specifically, the Corps has constructed—and continues to operate, maintain, and rehabilitate—a large inventory and wide variety of water management infrastructure, including reservoirs, hydropower facilities, commercial inland waterways, harbors, and levee systems. In June 2011, in response to the implementing instructions for Executive Order 13514, the Corps established its adaptation policy statement for addressing the effects of climate change, which called for the integration of climate change adaptation in all Corps activities.[20] Two Corps programmatic efforts—the Interagency Performance Evaluation Task Force/Hurricane Protection Decision Chronology Lessons Learned Implementation Team (also known as Actions for Change) and the Responses to Climate Change Program—support the Corps' ongoing adaptation activities.

Since 1902, Reclamation has carried out its mission to manage, develop, and protect water and related resources in 17 western states. The agency has led or provided assistance in constructing most of the large dams and water diversion structures in the West for the purpose of developing water supplies for irrigation—that is, "reclaiming" these lands for human use. In 2009, Reclamation directed its Office of Research and Development and its Office of Policy and Administration to take the lead in implementing the actions required by the SECURE Water Act. Reclamation's climate adaptation efforts fall within the larger effort by Interior to, among other things, implement Executive Order 13514. In 2013, Interior adopted a policy directing the agency to integrate climate change adaptation strategies into its operations, policies, programs, and planning. See appendix I for more information on the Corps' and Reclamation's organization and infrastructure.

In addition to each agency's individual efforts, the Corps and Reclamation established a partnership in 2007—the Climate Change and Water Working Group (CCAWWG)—to address their mutual concerns about the potential effects of climate change on their agencies' missions.[21] In 2009, the CCAWWG partners jointly produced a preliminary assessment of how climate could impact federal water resources management, which explored strategies to improve water management by tracking, anticipating, and responding to climate change, and identified adaptation challenges.[22]

Agencies Manage Climate Change Risk While Facing Aging Infrastructure

According to a 2012 National Research Council report on Corps infrastructure,[23] large portions of the Corps' water resources infrastructure were built over 50 years ago and are experiencing various stages of decay and disrepair, making project maintenance and rehabilitation a high priority. The report also found that federal funding over the past 20 years has consistently been inadequate to maintain the Corps' infrastructure at acceptable levels of performance and efficiency.

Similarly, most of Reclamation's water infrastructure facilities are more than 50 years old and, according to a 2011 Congressional Research Service report, with limited budgetary resources and aging infrastructure, Reclamation's maintenance needs are likely to increase, as is competition for limited funding.[24]

Despite the ongoing challenges of operating and maintaining aging infrastructure under budgetary constraints, it is important that the Corps and Reclamation address the challenge of managing climate change risk in order to limit the fiscal exposure of the federal government.

As noted in our 2013 high-risk update, climate change poses a significant financial risk to the federal government, including, but not limited to its role as the owner or operator of extensive infrastructure vulnerable to climate impacts.

State and local authorities are responsible for planning and implementing many types of infrastructure projects, and decisions at these levels of government can affect the federal government's fiscal exposure.[25] While implementing adaptive strategies to protect infrastructure may be costly, there is a growing recognition that the cost of inaction could be greater and—given the government's precarious fiscal position—increasingly difficult to manage given expected budget pressures, which will constrain not just future ad hoc responses, but other federal programs as well.

As stated in a 2010 National Research Council report, increasing the nation's ability to respond to a changing climate can be viewed as an insurance policy against climate change risks.[26]

Efforts under Way by the Corps and Reclamation to Assess Water Resource and Infrastructure Vulnerabilities to Climate Change and Develop Adaptation Strategies

Since 2009, as directed by executive order or required by law, the Corps and Reclamation have taken steps to assess water resource and infrastructure vulnerabilities and develop guidance and strategies for adapting to climate change, as shown in table 2.[27] Officials from both agencies told us that as they develop the necessary guidance, they plan to implement specific adaptation strategies and share costs with state and local partners.

Table 1. Corps and Reclamation Efforts to Assess Climate Change Vulnerabilities and Develop Adaptation Strategies and Guidance

Activities	Corps	Reclamation
Assessing climate change vulnerabilities	Completed: High-level assessment of climate change impacts on agency missions and operations. Ongoing: Conducting an initial assessment of climate change impact vulnerability for coastal activities. Developing and testing methodologies for assessing climate impacts at the watershed level for inland activities and a separate methodology for coastal activities.	Completed: High-level assessment of climate change impacts on water supplies in the major river basins in the West. Ongoing: Assessment of climate change impacts to water demands, including agricultural demands.
Developing adaptation strategies and guidance	Ongoing: Conducting pilot studies to support the development of adaptation guidance and a portfolio of adaptation approaches. Developing guidance for considering adaptation in all agency activities and for the implementation of adaptation approaches.	Ongoing: Conducting more focused assessments within basins to develop strategies to mitigate climate change impacts on water supplies. Developing guidance for considering adaptation in specific agency activities.

Sources: GAO analysis of Corps and Reclamation information.

The Corps Completed a Mission-Level Vulnerability Assessment and Is Conducting Pilot Studies to Identify Adaptation Strategies

Since 2009, the Corps has broadly assessed how climate change could affect its missions. Specifically, a phased assessment of the vulnerability of coastal projects is under way, more refined watershed-level vulnerability assessments are being developed, and pilot studies are being conducted to develop adaptation guidance and strategies. In March 2012, responding to guidance for implementing Executive Order 13514, the Corps provided the President's Council on Environmental Quality with a high-level analysis of the vulnerability of the Corps' missions and operations to climate change.[28] The Corps' analysis included an assessment of whether the potential effects of climate change on the Corps' business areas would likely be negative, positive, or a mix of both.[29] The analysis found, for example, that increasing air temperatures may have an effect on glaciers that could negatively impact Corps business areas such as navigation, flood and coastal storm damage reduction, ecosystem restoration, and emergency management, but the effect of increasing air temperatures on river ice could have both positive and negative impacts on those same four business areas.

As part of the high-level analysis, the Corps also identified a number of adaptation priorities. To address its priority of developing more refined vulnerability assessments, the Corps is currently undertaking coastal project vulnerability assessments and is developing and testing a methodology for nationwide, watershed-level vulnerability assessments of its inland missions, operations, programs, and projects. The nationwide watershed-level vulnerability assessments will, according to Corps officials, help them make initial screening-level determinations of where adaptation strategies are needed or not needed, and then to prioritize accordingly. The agency officials expect to complete the first phase of screening assessments in 2013, with more refined assessments to come in future years. Ultimately, the Corps plans to combine the coastal and inland assessments into a unified methodology. In addition, Corps Civil Works officials told us that they are developing syntheses of literature to provide up-to-date information about regional climate impacts in order to support project-specific planning and help implement adaptation strategies.

As part of its efforts to address the adaptation priorities of developing a risk-informed decision-making framework for climate change adaptation and a portfolio of adaptation approaches, the Corps is conducting 15 pilot studies nationwide to test different methods and frameworks for adapting to climate change. The Corps has completed 5 of these studies and 10 others are ongoing.[30] According to Corps Civil Works officials, the pilot studies were proposed by Corps district staff and selected by senior Corps staff in a competitive, internal process. The 15 pilots are led by 13 different Corps districts and address project planning, engineering, operations, and maintenance for 6 different Corps business areas, involving a variety of different infrastructure types, such as flood risk reduction projects, reservoirs, and canals. Some of the pilots are being conducted collaboratively with federal, state, and university partners. Taken together, Corps officials expect the pilot projects to provide a body of knowledge and tested methods that will serve as the foundation for its guidance and future adaptation efforts. See appendix II for additional information on the locations and results of the Corps' pilot studies.

Reclamation Assessed Vulnerabilities in Major Western River Basins and Is Developing Potential Adaptation Strategies

Reclamation has broadly assessed how climate change may affect water resources in the western United States as part of the Basin Study Program it established to meet the requirements of the SECURE Water Act. Specifically, responding to the Act's requirement to assess specific risks to water supplies,[31] Reclamation reported to Congress in 2011 its assessments of the potential effects of climate change on water supplies in the major river basins in the West.[32] The studies, known as West-Wide Climate Risk Assessments (WWCRA), are high-level, baseline assessments of the potential impacts of climate change on future water supplies—including impacts on Reclamation's ability to deliver water and hydropower—for each of the major river basins where Reclamation owns and operates water management infrastructure. According to agency officials, Reclamation is now conducting WWCRAs that focus on future water demand and will combine this information with its water supply assessments to form a more complete picture of the potential impacts of climate change on its water infrastructure.[33] Reclamation policy officials told us that these combined assessments will be included in Reclamation's next SECURE Water Act report that is due in 2016. These assessments will also be updated based on the latest climate science as water uses and conditions change, and they will be included in future reports that are due to Congress every 5 years.[34]

In addition to its WWCRAs, Reclamation is partnering with nonfederal entities to conduct more focused assessments, known as Basin Studies, to identify specific water resources vulnerabilities and to implement the SECURE Water Act's requirement that Reclamation consider and develop strategies to mitigate climate change impacts on water supplies.[35] Reclamation selects the Basin Studies through a competitive process and shares their costs with nonfederal partners. Through the Basin Studies, Reclamation intends to identify basin-wide water supply vulnerabilities, project climate change's impacts on the performance of water infrastructure, and develop adaptation strategies—such as operational or physical changes to existing water infrastructure or development of new facilities—to address these impacts. According to Reclamation guidance, the Basin Studies are to develop long-term projections of water supply and demand that take into account specific climate change risks identified in the SECURE Water Act.[36] The studies will analyze how well existing water and power infrastructure are meeting current demands and then forecast their performance in light of projected water supply and demand. To address projected imbalances in supply and demand, the studies are to identify adaptation strategies that include strategies for nonstructural (i.e., management and operations) and structural (i.e., capital expenditures) changes.

As of September 2013, 3 Basin Studies have been completed, and an additional 14 studies have been funded and are under way.[37] Reclamation and its partners, including state water management agencies and local irrigation districts, completed the Yakima River Basin Study in 2011, and both the St. Mary River and Milk River Basins Study and Colorado River Basin Study in 2012. Some studies entirely cover the major river basins specified in the SECURE Water Act—such as the Colorado River Basin Study, while other studies cover subbasins or tributaries within the boundaries of the major river basins—such as the Yakima River Basin Study, covering a tributary of the Columbia River. Reclamation officials told us that they next intend to initiate feasibility studies for adaptation strategies identified in completed Basin Studies by making funds available to nonfederal partners, beginning with an initial feasibility study in 2013. See appendix II for additional information on the locations and results of Reclamation's Basin Studies.

Both Agencies Are Developing and Enhancing Guidance for Implementing Climate Adaptation Strategies Identified

As recommended by the Task Force in 2010, the Corps and Reclamation are taking a phased approach to climate adaptation, including developing agency-wide guidance for adaptation. Specifically, through broad climate vulnerability assessments, agency officials told us they have expanded their knowledge of climate change and its impacts, allowing them to assess, at a high level, how these impacts may affect agency missions, programs, and operations. These initial vulnerability assessments have informed the agencies in developing and conducting more detailed vulnerability assessments, while also identifying specific strategies for climate change adaptation through the pilots and basin studies. Both agencies have also begun integrating what they have learned into their policies and program guidance. For example, beginning in 2009 and updated in 2011, the Corps issued guidance requiring that potential sea- level changes be considered in all of the agency's coastal planning, engineering, operations, and maintenance activities. The Corps is currently developing guidance for implementing coastal and inland adaptation strategies.

Similarly, in 2013, Reclamation officials told us they began to incorporate climate change adaptation considerations into its policies and guidance for project feasibility studies and environmental impact studies, among other things, using information and lessons learned from its WWCRA and Basin Study vulnerability assessments.

According to agency officials, as the Corps and Reclamation integrate climate adaptation considerations into their policies and program guidance, they will begin to take steps toward implementing the potential adaptation strategies that they have identified. According to the Corps' 2012 adaptation plan, the agency's goal is to create a policy and guidance framework that will support the implementation of practical, nationally consistent, legally justifiable, and cost-effective adaptation strategies.

Accordingly, Corps Civil Works officials told us that the pilots conducted to date have been largely focused on informing the development of policy and guidance, and that no structural or operational adaptation strategies have been implemented. Similarly, Reclamation officials told us no structural or operational adaptation strategies have been implemented in response to the agency's vulnerability assessments, and that its efforts to update guidance will be informed by the adaptation strategy feasibility studies as they are completed.

Corps and Reclamation officials told us that because they are early in the adaptation process, the extent to which limited budgets and existing infrastructure maintenance backlogs will affect the implementation of adaptation strategies remains to be determined. However, the implementation of adaptation strategies by both agencies will likely rely on collaborative sharing of costs and resources with federal, state, local, and nongovernmental stakeholders.

Both agencies have already initiated cost-sharing and resource-leveraging measures. For example, according to agency officials, the Corps is leveraging resources for adaptation pilot studies with state, local, and nongovernmental entities, and Reclamation is splitting the cost of Basin Studies with state and local partners. The agencies plan to continue such collaborative approaches going forward. For example, as required under the SECURE Water Act, Reclamation officials told us they intend to share the cost of the feasibility studies for adaptation strategies equally with nonfederal partners.

THE CORPS AND RECLAMATION FACE SEVERAL CHALLENGES IN ASSESSING AND RESPONDING TO THE EFFECTS OF CLIMATE CHANGE AND ARE WORKING COLLABORATIVELY TO ADDRESS THEM

In 2009, the Corps and Reclamation—with its CCAWWG partners NOAA and USGS—published a study,[38] (referred to in this report as the CCAWWG study). This study identified several challenges that climate change poses for water resource managers, including (1) identifying the data and tools needed by water managers to address climate change, (2) ensuring the sustained collection of climate data, (3) incorporating climate science into water management tools, and (4) educating water managers to use climate data and tools.

We found that the Corps and Reclamation are addressing these challenges, making collaboration a key element of their efforts, and doing so in a manner generally consistent with best practices for sustained collaboration.

The Corps and Reclamation Are Addressing Several Challenges to Climate Adaptation

The CCAWWG study identified a number of challenges faced by the Corps and Reclamation in adapting to climate change, and the agencies have taken a variety of actions to address these challenges.

- *Identifying the data and tools needed by water managers to address climate change:* The CCAWWG agencies are collaborating to produce a series of four documents identifying their common data and tool needs and a strategy for meeting them, with the objective, among other things, of guiding and fostering federal and nonfederal research and technology investments toward meeting these needs.[39] The first document, published in 2011, described the water management community's needs for climate change information and tools to support long-term planning for time scales of 5 years and more. The second document, published in 2013, described the data and tools needed to support short-term planning of less than 5 years. For both documents, the Corps and Reclamation asked their water resource managers to identify the information and tools most relevant to their programs, and they also consulted with other federal, state, and local agencies and stakeholders with a role in water resource management about their needs. The documents summarized the information and tools needed into categories and identified the users' most pressing needs within each category. According to these documents, the CCAWWG agencies plan next to prepare two companion documents to identify a scientific strategy for meeting the research needs identified in the two initial documents. The two completed documents note that USGS and NOAA will jointly prepare the companion documents, incorporating perspectives from other federal and nonfederal representatives of the scientific community.
- *Ensuring the sustained collection of climate data:* The Corps and Reclamation are coordinating with the data collecting agencies and sharing some costs associated with their efforts. According to the 2009 CCAWWG study, at the same time as the need for observational data to support climate adaptation is increasing, the observational networks crucial to increasing understanding are shrinking. For example, in recent decades, maintenance of long-term monitoring networks has declined because of a lack of funding—USGS alone has

deactivated or discontinued almost 1,700 surface-water stream gauges. Corps Civil Works officials told us that stream gauge data is extremely important, not only to the Corps' ongoing operations, but also because science agencies use the data to produce the climate change information upon which the Corps bases its adaptation planning. To ensure the needed data is collected, the Corps has a formal agreement to provide funding to USGS—about $18 million in fiscal year 2013 according to Corps officials—to operate stream gauges that provide data for the Corps' water planning and management activities. This data is also available for use by all federal and state agencies, as well as others interested in water information.

Reclamation policy officials told us, in response to a SECURE Water Act requirement to consult with federal and applicable state agencies to develop a monitoring plan for acquiring and maintaining water resources information, agency staff are currently identifying information needs and plans to work collaboratively with data collecting agencies, including NOAA, USGS, and the U.S. Department of Agriculture, to develop the plan.[40] Reclamation policy officials stated that the Basin Study Program activities, including the WWCRAs and Basin Studies, are also contributing to the planning effort by providing valuable information about how the agency's monitoring needs are changing as a result of climate change. Reclamation officials told us that they intend to work with the U.S. Department of Agriculture and USGS to initiate the required monitoring plan in 2014. These collaborative efforts can help ensure that the long-term water resource monitoring networks critical for detecting and quantifying climate change and its impacts—as well as measuring the effectiveness of future adaptation strategies—will be properly configured and continue to operate.

- *Incorporating climate science into water management tools:* The Corps, Reclamation, and others are collaborating in a number of efforts to incorporate climate science into water management tools.[41] For example, Reclamation research and development, as well as policy officials told us that as part of their effort to enhance the capabilities of water resource managers to use climate data, the agency is coleading two of Interior's Landscape Conservation Cooperatives in the Colorado River Basin area. Reclamation officials told us that the Landscape Conservation Cooperatives, which are partnerships of governmental and nongovernmental stakeholders, will focus on developing and communicating science to inform climate adaptation strategies for ecological regions, or "landscapes." In collaboration with academia, other federal agencies, local and state partners, and the public, the Landscape Conservation Cooperatives will provide products and services, such as climate change computer models and vulnerability assessments, coordinate with Interior's regional Climate Science Centers to synthesize existing climate change impact data and management strategies, and help resource managers put them into action on the ground.[42] The Landscape Conservation Cooperatives will also coordinate with NOAA's Regional Integrated Sciences and Assessment program.[43]

Similarly, the Corps is engaged in collaborative efforts with external partners to integrate climate science into planning tools for water resource managers. For example, the Corps partnered with NOAA's National Ocean Service to create an online sea level change calculator. According to Corps officials, this collaboration allowed the rapid integration of climate science into engineering guidance for coastal projects. Corps Civil Works officials also told us that collaboration with the Federal Emergency Management Agency and NOAA's Urban Northeast Regional Integrated Sciences and Assessment program contributed to the development of a post-Superstorm Sandy sea level rise tool to help affected communities, residents, and other stakeholders consider risks from future sea level rise in planning for reconstruction.[44]

In addition, the Corps and NOAA are coleading actions to implement the Task Force recommendation to develop a federal Internet portal to provide current, relevant, and high-quality information on water resources and climate change data applications and tools for assessing the vulnerability of water programs and facilities to climate change. As a result, the Corps-hosted Federal Support Toolbox is now publicly available online. According to Corps Civil Works officials, the website is a "one stop shop" for technical resources to support water management. The website states that the Toolbox is an evolving and comprehensive water resources data portal with direct links to valuable databases, innovative programs and initiatives, and state-of-the-art models and tools.[45]

- *Educating water managers to use climate data and tools:* The Corps and Reclamation are collaborating as coleaders in developing and implementing a training program as recommended by the Task Force's *National Action Plan: Priorities for Managing Freshwater Resources in a Changing Environment.*[46] These agencies—joined by NOAA, USGS, and the Environmental Protection Agency—are collaborating with the University Corporation for Atmospheric Research's COMET Program and the Western Water Assessment to produce training courses for federal and nonfederal resource management professionals who need to assess the impacts of climate change on water and related resources.[47] Specifically, in 2012, the agencies implemented an online course on incorporating climate change into water resource planning, which was a prerequisite for participating in the first two pilot residence courses—addressing climate impacts on surface hydrology and on water demand for irrigated crops—offered in early 2013. According to the program's website, the online course was designed to provide students with water resource planning knowledge, while the residence courses offer opportunities for gaining hands-on experience in applying that knowledge. According to the website, these courses and the numerous planned future classes are collectively designed to provide a professional development and training series that will help managers assess climate change impacts across the spectrum of natural resources.

The Corps' and Reclamation's Collaborative Efforts Are Generally Consistent with the Best Practices for Enhanced and Sustained Collaboration

The Corps and Reclamation have collaborated with each other and other agencies in a manner that is generally consistent with practices that we have previously identified as important to helping enhance and sustain collaboration among federal agencies.[48] In 2005, we reported that collaboration—broadly defined as any joint activity that is intended to produce more public value than could be produced when organizations act alone—can be enhanced and sustained by engaging in eight key practices: (1) defining and articulating a common outcome; (2) establishing mutually reinforcing or joint strategies; (3) identifying and addressing needs by leveraging resources; (4) agreeing on roles and responsibilities; (5) establishing compatible policies, procedures, and other means to operate across agency boundaries; (6) developing mechanisms to monitor, evaluate, and report on results; (7) reinforcing agency accountability for collaborative efforts through agency plans and reports; and (8) reinforcing individual accountability for collaborative efforts through performance management systems. Running throughout these practices are a number of factors, such as leadership, trust, and organizational culture, which are necessary elements for a collaborative working relationship.

The Corps and Reclamation have made collaboration a key element of their adaptation policies and plans and have reinforced accountability for collaboration through agency performance management systems. For example, the Corps' climate adaptation policy states that collaborations are the most effective way to develop strategies to identify and reduce vulnerabilities to potential future climate change, and it calls for continued collaborative adaptation efforts. As stated in the Corps' 2012 *Climate Adaptation Plan and Report*, it is the objective of the agency to facilitate and promote closer and more fruitful interagency cooperation and to promote sharing of impact and adaptation data and information between federal, state, and local partners. Finally, to reinforce accountability, the draft performance metrics for climate adaptation in the Corps' 2013 *Army Campaign Plan* includes a target for the number of products developed in collaboration with other water resource agencies for adaptation planning and action.

Reclamation has similarly included collaboration as a key element of its adaptation policy, plans, and performance metrics. Under Interior's 2013 Climate Change Adaptation Policy, Reclamation is to integrate climate change adaptation strategies into its policies and practices by, among other actions, collaborating with stakeholders through Landscape Conservation Cooperatives, Climate Science Centers, and other partnerships to increase understanding of climate change. Furthermore, Reclamation's strategic plan for its adaptation and conservation programs states that collaborative partnerships must be developed to identify the adaptive strategies needed to address climate change. Finally, through its climate adaptation efforts, Reclamation is contributing to Interior's goal of identifying resources that are particularly vulnerable to climate change and implementing coordinated adaptation responses for half of the nation by September 30, 2013.

We also found that the Corps and Reclamation have collaborated together and with others in accordance with best practices for collaboration among agencies. For example, in their key collaborative effort—CCAWWG—each agency's role is well defined; the Corps and Reclamation provide water engineering and management expertise, and their partner agencies provide climate science expertise. CCAWWG has clearly defined common objectives, including the development of working level relationships between federal water management and federal science agencies, and it leverages resources across agencies to meet common needs. The CCAWWG agencies also have mutually reinforcing strategies. For example, the operating needs of the Corps and Reclamation drive the direction of science inquiries by the science agencies, resulting in improved operations, while at the same time the data collected and compiled by the water management agencies for a specific purpose can be used by the science agencies for alternative objectives. Outside of CCAWWG, as mentioned elsewhere in this report, the agencies have also followed key collaborative practices, such as the Corps leveraging resources to fund maintenance of the USGS's stream-flow monitoring networks, and Reclamation establishing joint strategies with state agencies and others to conduct Basin Studies.

AGENCY COMMENTS

We provided a draft of this product for review and comment to the Departments of Defense and the Interior. The Department of Defense provided technical comments that were incorporated, as appropriate.

Steve D. Morris
Acting Director, Natural Resources and Environment

APPENDIX I: THE U.S. ARMY CORPS OF ENGINEERS' AND THE BUREAU OF RECLAMATION'S ORGANIZATION AND INFRASTRUCTURE

The U.S. Army Corps of Engineers (Corps)—an agency within the Department of Defense and a Major Command within the Army—is composed of four program areas including Civil Works, Military Construction, Real Estate, and Research and Development. To carry out its Civil Works missions nationwide, the program is organized into eight geographic divisions composed of 38 districts, as shown in figure 1. Division and district geographic boundaries are generally aligned with watershed boundaries.

Source: U.S. Army Corps of Engineers, Map Resources (map).

Note: In addition to the eight divisions that manage U.S. water resources, the Corps also operates a Gulf Region Division, Afghanistan Engineer District, and Europe District.

Figure 1. U.S. Army Corps of Engineers Division and District Boundaries.

The Corps' Civil Works Program is implemented through nine business areas that represent the diversity of the nation's water resource management needs as follows:

- Navigation—provides safe and reliable commercial waterways;
- Flood and Coastal Storm Damage Reduction—reduces risks to people, homes, and communities from flooding and coastal storms;

- Environment—restores and protects wetlands and other aquatic ecosystems;
- Hydropower—generates hydroelectric power for distribution to communities;
- Regulatory—regulates work in navigable rivers and the discharge of dredges and fill materials in U.S. waters;
- Recreation—provides recreational and educational opportunities;
- Emergency Management—prepares for natural disasters and acts when they occur;
- Water Supply—provides water storage for multiple purposes; and
- Executive Direction and Management—provides leadership, strategic planning and performance measurement.

To implement its responsibilities in the business areas, the Corps has constructed—and continues to operate, maintain, and rehabilitate—a large inventory and wide variety of water management infrastructure. For example, according to Corps records, as of September 2012, its inventory includes 702 reservoirs that, among other things, provide water supply storage and help reduce flood risk, and 75 hydropower facilities with 353 generating units that produce hydroelectric power for homes, businesses, and communities. In the area of navigation, the Corps' inventory includes 12,000 miles of commercial inland waterways, 193 lock sites with 239 chambers, and 926 harbors.[49] For flood reduction, the Corps inventory includes 14,501 miles of levee systems. For recreation, the Corps' inventory includes 54,879 miles of lakeshore and recreation areas that support 370 million annual visitors, 270,000 jobs, and $16 billion of economic activity. In 2012, the National Research Council reported that the Corps' infrastructure had an estimated value of approximately $164 billion.

The Department of the Interior's Bureau of Reclamation (Reclamation) is, according to agency records, the nation's largest wholesale water supplier, providing water to over 31 million people and irrigation water to one out of five western farmers on 10 million acres of farm land, which produce 60 percent of the nation's vegetables and one-quarter of its fresh fruit and nut crops. Reclamation is also the second largest producer of hydropower in the United States. To carry out its mission and operations, Reclamation is organized into five regions, as shown in figure 1.

Sources: U.S. Bureau of Reclamation, Map Resources (map).

Figure 2. Bureau of Reclamation Regional Boundaries.

According to agency records as of 2011, Reclamation's asset inventory includes 476 dams and dikes, creating 337 reservoirs with a total storage capacity of 245 million acre-feet of water.[50] Reclamation's inventory also includes 53 hydroelectric power plants that it owns and operates. These plants provide an average of more than 40 megawatt hours of energy per year.[51] In 2011, Reclamation, estimated that the replacement value of its assets was approximately $90 billion. According to the Congressional Research Service, about two-thirds of the assets in Reclamation's inventory are "transferred works"—facilities that it owns, but for which it has transferred operations and maintenance to nonfederal entities. The remaining one-third of these facilities are "reserved works"—facilities that are owned, operated, and maintained by Reclamation, where the nonfederal beneficiaries, such as irrigation districts, are responsible for repaying construction and maintenance costs.

APPENDIX II: LOCATIONS AND INITIAL RESULTS OF THE U.S. ARMY CORPS OF ENGINEER'S PILOT STUDIES AND BUREAU OF RECLAMATION'S BASIN STUDIES

The Corps is conducting pilot studies nationwide to support the development of adaptation guidance and a portfolio of adaptation approaches. Reclamation is conducting Basin Studies—focused assessments within basins—to develop strategies to mitigate climate change impacts on water supplies.

Corps Pilot Studies

As part of its efforts to address the priorities of developing a risk-informed decision-making framework and a portfolio of adaptation approaches, the Corps is conducting 15 pilot studies nationwide to test different methods and frameworks for adapting to climate change (see Figure 3). As of September 2013, the Corps has completed 5 of these studies, and 10 others are ongoing. The Corps plans to initiate 1 additional pilot in 2013.

According to the Corps' September 2012 report on its climate change adaptation pilots,[52] each study addresses a central question designed to help the Corps, according to officials, test new ideas, develop and utilize information at the project-level scale, and to collect information needed to develop policy and guidance for incorporating adaptation into all agency activities. For example, because the reliability of a planned flood reduction project could depend on how climate change affects future flooding, the Corps launched a pilot study with the goals of determining (1) whether tools and data are available for the Corps to provide reliable estimates of future flooding using climate projections and (2) how changes in precipitation patterns will affect flood events. This pilot study's results supports the Corps' effort to design an approximately $1.8 billion flood risk reduction project on the Red River of the North, where flooding has increased in magnitude and frequency since 1942. According to

Corps officials, the pilot study successfully adapted existing tools and data to project future flooding, while also testing current guidance and contributing to the development of new guidance for future adaptation efforts. Specifically, Corps officials told us that climate data indicated a trend of flooding consistent with climate change projections. In accordance with existing Corps risk analysis guidance, the pilot study leader convened an expert panel to review the climate information. Based on the expert panel's findings, the pilot team is integrating climate data into the Corps' flood model to project future river conditions. In addition, a member of the pilot study team who is also responsible for developing the Corps' guidance for inland adaptation activities told us that he is integrating his observations and experiences from this study—as well as from his participation in other Corps pilots—into the guidance.

Corps officials told us that the information developed by the Corps' climate change adaptation pilots is also contributing to subsequent Corps studies. For example, the climate change and modeling data for a pilot study of sediment impacts to Cochiti Dam and Lake in New Mexico is contributing to ongoing studies in the Corps' Albuquerque District. One is the Santa Clara Pueblo Watershed Assessment, which is studying observed climate trends and projected climate changes to address likely future changes to watershed hydrology on the Pueblo's lands, with particular attention to flood risk and water resources development.

Reclamation Basin Studies

Reclamation is partnering with nonfederal entities to conduct focused assessments, known as Basin Studies, to identify specific water resources vulnerabilities and to implement the SECURE Water Act's requirement that Reclamation consider and develop strategies to mitigate climate change impacts on water supplies.[53] As of September 2013, 3 Basin Studies have been completed, and an additional 14 studies have been funded and are under way in western states (see Figure 4).[54]

All three of Reclamation's completed Basin Studies projected climate change to result in warming and changes in precipitation that will alter the snow cover and runoff that supply water to the river basins, although the changes may be difficult to predict with certainty for any particular location and time. For example, the Colorado River Basin study projected that snow cover will decrease in the basin as more precipitation falls as rain, rather than snow, and warmer temperatures cause earlier melting.

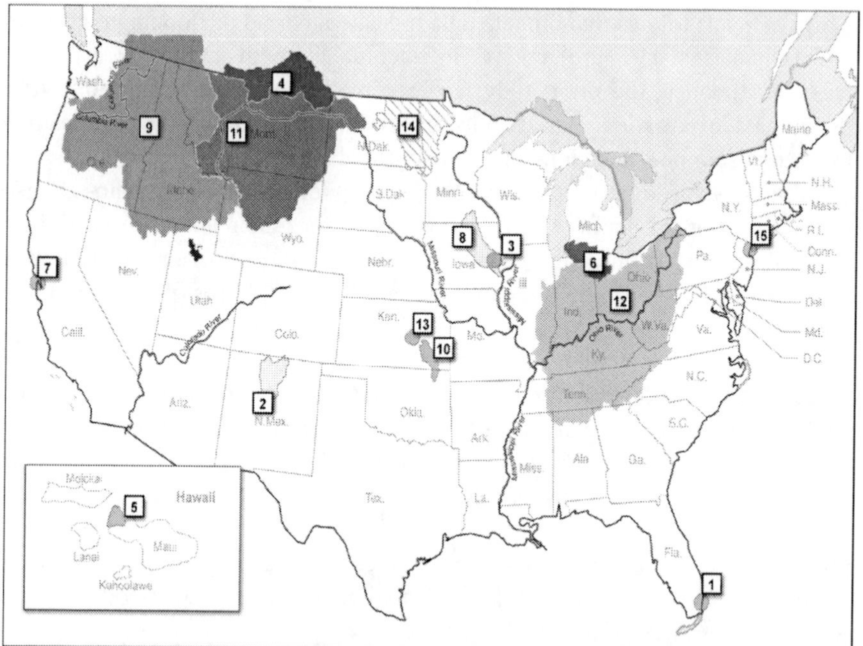

1 Application of Sea-Level Change Guidance to C-111 Spreader Canal, Florida
2 Climate Change Associated Sediment Yield Impacts on the Rio Grande, Cochiti Dam and Lake
3 Climate Change Impacts on the Operation of Coralville Lake, Iowa
4 Climate Change Associated Sediment Yield Impacts and Operation Evaluations at Garrison Dam, North Dakota
5 Risk-Informed Decision-Making for Integrated Water Resource Management Planning, West Maui Watershed
6 Upland Sediment Production and Delivery in the Great Lakes Region under Climate Change
7 Risk-Informed Decision-Making for Potential Sea-Level Rise Impacts on the Hamilton Wetland Restoration Project, California
8 Climate Modeling and Stakeholder Engagement to Support Adaptation in the Iowa-Cedar Watershed
9 Framework for Building Resiliency into Restoration Planning – Lower Columbia River Estuary
10 Climate Change Impacts on Water Supply in Marion Reservoir Watershed, Kansas
11 Missouri River Basin Mountain Snowpack – Accumulation and Runoff
12 Formulating Mitigation/Adaptation Strategies through Regional Collaboration with the Ohio River Basin Alliance
13 Utilization of Regional Climate Science Programs in Reservoir and Watershed Risk-Based Assessments, Oologah Lake and Watershed
14 Red River of the North Flooding at Fargo, North Dakota
15 East Rockaway Inlet to Rockaway Inlet, New York, Collaboration Framework Development

Sources: U.S. Army Corps of Engineers, Map Resources (map).
Note: Pilot studies 1-5 have been completed as of September 2013.

Figure 3. Location and Description of U.S. Army Corps of Engineers Pilot Projects, as of September 2013.

As a result, the runoff in the basin is generally projected to decrease—except in the northern Rockies. The study concluded that without additional water management actions, a wide range of future imbalances in supply and demand are plausible, primarily because of the uncertainty in future water supplies.

The completed Basin Studies also found that climate change is expected to contribute to long-term imbalances in water supply and demand. The completed Basin Studies also identified, with varying degrees of specificity, possible adaptation strategies to address the projected effects of climate change on water resources.

For example, the Colorado River Basin Study identified various combinations of potential adaptation strategies to address projected water supply and demand imbalances.

The study noted that the purpose of identifying these strategies was not to select a single, best strategy, but rather to recognize that there are various ways to address projected water imbalances in the basin, and that each approach has positive and negative implications that should be considered in future planning efforts.

Agency officials and study participants told us that Reclamation has formed workgroups to further refine the strategies and identify options that should then be evaluated through feasibility studies. In contrast, the Yakima River Basin Study identified a more specific set of adaptation strategies for which feasibility studies are planned. Nonfederal stakeholders, including those representing agricultural and the environmental interests, told us they are collaboratively pursuing state funding to initiate the feasibility studies.

Reclamation policy officials and several other basin stakeholders—including state water managers, environmental advocates, and local water providers—told us that a key outcome of the completed Basins Studies is the establishment of a shared view of how climate change will impact the basins and their water management infrastructure.

Reclamation policy officials, as well as basin stakeholders representing environmental and tribal interests, told us that the Basin Studies represent Reclamation's first concerted efforts at holistic planning in the river basins, taking into account not only the needs and concerns of irrigation users, but also the interests of tribes, recreational users, environmental advocates, and others. A state water manager told us that a prior river basin planning effort had largely been unsuccessful due to the lack of involvement or opposition by environmental and tribal interests, and several stakeholders commended Reclamation for working to ensure that the views of a wide range of stakeholders, including environmental and tribal interests, were considered in the Basin Studies.

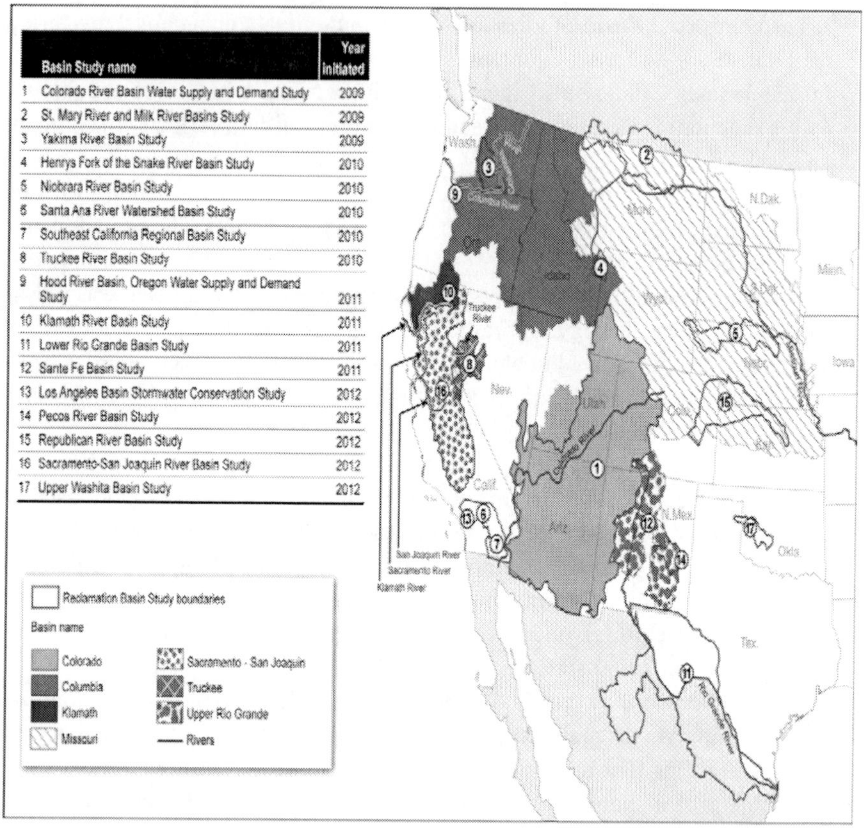

	Basin Study name	Year Initiated
1	Colorado River Basin Water Supply and Demand Study	2009
2	St. Mary River and Milk River Basins Study	2009
3	Yakima River Basin Study	2009
4	Henrys Fork of the Snake River Basin Study	2010
5	Niobrara River Basin Study	2010
6	Santa Ana River Watershed Basin Study	2010
7	Southeast California Regional Basin Study	2010
8	Truckee River Basin Study	2010
9	Hood River Basin, Oregon Water Supply and Demand Study	2011
10	Klamath River Basin Study	2011
11	Lower Rio Grande Basin Study	2011
12	Sante Fe Basin Study	2011
13	Los Angeles Basin Stormwater Conservation Study	2012
14	Pecos River Basin Study	2012
15	Republican River Basin Study	2012
16	Sacramento-San Joaquin River Basin Study	2012
17	Upper Washita Basin Study	2012

Sources: Bureau of Reclamation, Map Resources (map).

Figure 4. Locations of the Bureau of Reclamation's 17 Funded Basin Studies, as of September 2013.

End Notes

[1] The U.S. Global Change Research Program coordinates and integrates federal research on changes in the global environment and their implications for society. Led by a team of officials from each of the U.S. Global Change Research Program's 13 participating departments and agencies, the U.S. Global Change Research Program engages in a variety of activities designed to strengthen and strategically direct climate change research in the United States and improve the flow of that information to federal, state, and local decision makers, and the public.

[2] Thomas R. Karl, Jerry M. Melillo, and Thomas C. Peterson, eds., U.S. Global Change Research Program, *Global Climate Change Impacts in the United States* (New York, New York: Cambridge University Press, 2009).

[3] Lins, Harry F., Hirsch, Robert M., and Kiang, Julie, *Water—the Nation's Fundamental Climate Issue: A White Paper on the U.S. Geological Survey Role and Capabilities: U.S. Geological Survey Circular 1347 (2010).*

[4] GAO, *Climate Change Adaptation: Strategic Federal Planning Could Help Government Officials Make More Informed Decisions,* GAO-10-113 (Washington, D.C.: Oct. 7, 2009).

[5] *National Research Council, America's Climate Choices: Adapting to the Impacts of Climate Change* (Washington, D.C.: 2010).

[6] GAO, *High-Risk Series: An Update*, GAO-13-283 (Washington, D.C.: February 2013).

[7] Pub. L. No. 111-11, tit. IX, subtit. F (2009).

[8] Executive Order 13514, *Federal Leadership in Environmental, Energy, and Economic Performance*, October 5, 2009. 74 Fed. Reg. 52117 (Oct. 5, 2009).

[9] *Instructions for Implementing Climate Change Adaptation Planning in Accordance with Executive Order 13514, Federal Leadership in Environmental, Energy, and Economic Performance*, March 4, 2011.

[10] GAO, *Results-Oriented Government: Practices That Can Help Enhance and Sustain Collaboration among Federal Agencies*, GAO-06-15 (Washington, D.C.: Oct. 21, 2005).

[11] Federal Interagency Panel on Climate Change and Water Data and Information, 2011, *Report to Congress—Strengthening the scientific understanding of climate change impacts on freshwater resources of the United States*, 49.

[12] GAO, *Climate Change Various Efforts Are Under Way at Key Natural Resources Management Agencies,* GAO-13-253 (Washington, D.C.: May 31, 2013).

[13] L.D. Brekke, J.E. Kiang, J.R. Olsen, R.S. Pulwarty, D.A. Raff, D.P. Turnipseed, R.S. Webb, and K.D. White, *Climate change and water resources management—A federal perspective: USGS Circular 1331 (2009).*

[14] Pub. L. No. 111-11, tit. IX, subtit. F (2009).

[15] In addition, the law requires Reclamation to (1) coordinate with USGS, NOAA, and appropriate state water resource agencies to ensure that its climate change program has access to the best available scientific information on presently observed and projected future impacts of global climate change on water resources; (2) consult with other federal and applicable state agencies to develop a monitoring plan to acquire and maintain water resources information; and (3) submit a report to relevant congressional committees that describes, among other things, the results of its assessments and analysis, the mitigation strategies developed and implementation of the monitoring plans within 2 years of its passage, and every 5 years thereafter.

[16] In our 2013 High-Risk Update, we noted that while individual agency actions are necessary, a centralized strategy driven by a government-wide plan is also needed to reduce the federal fiscal exposure to climate change, maximize investments, achieve efficiencies, and better position the government for success. Even then, such approaches will not be fully sufficient unless also coordinated with decisions at the state and local levels that drive much of the federal government's fiscal exposure. The challenge is to develop a cohesive approach at the federal level that also informs action at the state and local levels.

[17] Council on Environmental Quality, *Progress Report of the Interagency Climate Change Adaptation Task Force: Recommended Actions in Support of a National Climate Change Adaptation Strategy*, October 5, 2010.

[18] Council on Environmental Quality, *Federal Agency Climate Change Adaptation Planning, Implementing Instructions* and *Federal Agency Climate Change Adaptation Planning, Support Document,* March 4, 2011.

[19] In addition, as recommended in the 2010 progress report, the Task Force issued the *National Action Plan—Priorities for Managing Freshwater Resources in a Changing Climate* in October 2011, which established a goal for managing water resources in a changing climate, identified key recommendations to achieve that goal, and described specific supporting actions that federal agencies should take to implement the recommendations.

[20] The *USACE Climate Change Adaptation Policy Statement* issued June 3, 2011, established a Climate Change Adaptation Steering Committee to oversee and coordinate agency-wide climate change adaptation planning and implementation. The Corps' Climate Change Adaptation Steering Committee is chaired by the Chief, Engineering and Construction. The Assistant Secretary of the Army for Civil Works was designated as the USACE Senior Adaptation Point of Contact responsible for ensuring implementation of the Corps' adaptation policy.

[21] In addition to the Corps and Reclamation, founding CCAWWG members included USGS and NOAA—two key climate science agencies. CCAWWG has since expanded to include, among others, the U.S. Environmental Protection Agency, Federal Emergency Management Agency, National Aeronautics and Space Administration, and U.S. Department of Agriculture.

[22] L.D. Brekke, J.E. Kiang, J.R. Olsen, R.S. Pulwarty, D.A. Raff, D.P. Turnipseed, R.S. Webb, and K.D. White, *Climate change and water resources management—A federal perspective: USGS Circular 1331(2009).*

[23] Committee on U.S. Army Corps of Engineers Water Resources Science, Engineering, and Planning; Water Science and Technology Board; Division on Earth and Life Studies; National Research Council, *Corps of Engineers Water Resources Infrastructure: Deterioration, Investment, or Divestment?* (The National Academies Press, Washington, D.C.: 2012).

[24] Congressional Research Service, *The Bureau of Reclamation's Aging Infrastructure,* March 30, 2011.

[25] The federal government annually invests billions of dollars in infrastructure projects that state and local governments prioritize and supervise through, for example, decisions on zoning and how to build vulnerable infrastructure such as roads and bridges. See GAO-10-113.

[26] National Research Council, *America's Climate Choices: Adapting to the Impacts of Climate Change* (Washington, D.C.: 2010).

[27] Other activities to address adaptation challenges, such as the development of new climate and hydrologic projections, computer tools to work with the projections, websites to facilitate user access to these resources, and training programs for on-the-ground water resources practitioners, are discussed later in this report.

[28] The Corps included excerpts of this assessment in its June 2012 *Climate Adaptation Plan and Report*, provided to the Council on Environmental Quality as required by Executive Order 13514.

[29] Corps business areas cover the following: navigation, flood and coastal storm damage reduction, environment, hydropower, regulatory, recreation, emergency management, and water supply.

[30] The Corps plans to initiate one additional pilot in 2013.

[31] Pub. L. No. 111-11, § 9503(b)(2) (2009).

[32] The SECURE Water Act identifies eight major reclamation river basins, but Reclamation only conducted seven West-Wide Climate Risk Assessments because one covers both the Sacramento and San Joaquin River Basins.

[33] Reclamation officials told us the assessments are meant to fulfill the requirements of section 9503(b)(3) of the SECURE Water Act.

[34] Pub. L. No. 111-11, § 9503(c) (2009).

[35] Pub. L. No. 111-11, § 9503(b)(4) (2009).

[36] Pub. L. No. 111-11, § 9503(b)(2) (2009).

[37] Two additional studies were selected for funding in fiscal year 2013.

[38] L.D. Brekke, J.E. Kiang, J.R. Olsen, R.S. Pulwarty, D.A. Raff, D.P. Turnipseed, R.S. Webb, and K.D. White, *Climate change and water resources management—A federal perspective: USGS Circular 1331 (2009).*

[39] *Addressing Climate Change in Long-Term Water Resources Planning and Management: User Needs for Improving Tools and Information,* U.S. Army Corps of Engineers and U.S.

Department of the Interior, Bureau of Reclamation (Washington D.C.: January 2011) and *Short-Term Water Management Decisions: User Needs for Improved Climate, Weather, and Hydrologic Information,* U.S. Army Corps of Engineers, Bureau of Reclamation, and NOAA (Washington D.C.: January 2013).

[40] We found in previous reports that gaps in NOAA's satellite coverage, which could occur as soon as 2014, are expected to affect the continuity of climate and space weather measurements. Given the importance of satellite data to weather forecasts, the likelihood of significant gaps, and the potential impact of such gaps on the health and safety of the U.S. population and economy, GAO has concluded that the potential gap in weather satellite data is a high-risk area and added it to the High-Risk List in 2013. For example, see GAO, *Environmental Satellites: Focused Attention Needed to Mitigate Program Risks*, GAO-12-841T (Washington, D.C.: June 27, 2012). See also GAO, *Environmental Satellites: Strategy Needed to Sustain Critical Climate and Space Weather Measurements*, GAO-10-456 (Washington, D.C.: Apr. 27, 2010).

[41] In our 2013 High-Risk Update, we identified the need to develop a government-wide approach for providing (1) the best available climate-related data for making decisions at the state and local level and (2) assistance for translating available climate-related data into information that officials need to make decisions.

[42] Interior's eight regional Climate Science Centers deliver basic climate change impact science to Landscape Conservation Cooperatives within their respective regions, including physical and biological research, ecological forecasting, and multiscale modeling. The centers prioritize their delivery of fundamental science, data, and decision-support activities to meet the needs of the Landscape Conservation Cooperatives

[43] The Regional Integrated Sciences and Assessment program consists of 11 research teams that emphasize collaborative regional climate research to help inform climate change adaptation planning and management.

[44] In 2012, Superstorm Sandy caused tens of billions of dollars in damages to buildings, utilities, transportation systems, and other infrastructure. As we pointed out in our 2013 High-Risk Update, as weather-related events become more frequent and intense due to climate change, whatever resulting damages are not covered by insurance or built to be resilient to such events increases the federal government's implicit fiscal exposure through federal disaster relief programs.

[45] The Federal Support Toolbox is available at www.watertoolbox.us

[46] According to the CCAWWG study, adaptation to climate change will likely require water resource managers to move away from the traditional planning assumption of stationarity—the idea that historical records of stream flow and weather variation represent a reasonable proxy for the future—in designing and operating water resource systems. In the absence of stationarity, water managers will need to use new methods to estimate the likelihood of future hydrologic conditions, including approaches that use climate models to project a range of possible future condition scenarios.

[47] The University Corporation for Atmospheric Research is a consortium of over 100 member universities and academic affiliates focused on research and training in the atmospheric and related earth system sciences. Its COMET Program uses innovative methods to disseminate and enhance scientific knowledge in the environmental sciences, particularly meteorology, but also including diverse areas such as oceanography, hydrology, space weather, and emergency management. The Western Water Assessment is a university-based applied research program that addresses societal vulnerabilities related to climate, particularly in the area of water resources. Its mission is to identify and characterize regional vulnerabilities to and impacts of climate variability and change and to develop information, products, and processes to assist decision makers throughout Colorado, Utah, and Wyoming.

[48] GAO-06-15.

[49] A lock is a navigation device with one or more fixed chambers in which the water level can be varied, allowing the raising and lowering of boats between stretches of water of different levels on rivers and canals.

[50] An acre-foot is 325,851 gallons of water that, according to Reclamation, supplies enough water for a family of four for 1 year.

[51] A megawatt hour is one thousand kilowatt-hours or one million watt-hours, which will power the average American home for 1 week.

[52] U.S. Army Corps of Engineers, *Climate Change Adaptation Pilots* (June 2012).

[53] Pub. L. No. 111-11, § 9503(b)(4) (2009).

[54] Two additional studies were selected for funding in fiscal year 2013: the San Diego Watershed Basin Study and the West Salt River Valley Basin Study. Reclamation is developing the scopes of the studies with the respective funding partners.

In: Climate Change
Editor: Aileen Stablum

ISBN: 978-1-63117-584-8

Chapter 2

WATER—THE NATION'S FUNDAMENTAL CLIMATE ISSUE: A WHITE PAPER ON THE U.S. GEOLOGICAL SURVEY ROLE AND CAPABILITIES*

Harry F. Lins, Robert M. Hirsch and Julie Kiang

CONVERSION FACTORS

Multiply	By	To obtain
	Length	
inch (in.)	2.54	centimeter (cm)
inch (in.)	25.4	millimeter (mm)
foot (ft)	0.3048	meter (m)
	Flow rate	
cubic foot per second (ft^3/s)	0.02832	cubic meter per second (m^3/s)
cubic foot per second per square mile [(ft^3/s)/mi^2]	0.01093	cubic meter per second per square kilometer [(m^3/s)/km^2]
cubic foot per day (ft^3/d)	0.02832	cubic meter per day (m^3/d)
inch per hour (in/h)	0 .0254	meter per hour (m/h)
inch per year (in/yr)	25.4	millimeter per year (mm/yr)
mile per hour (mi/h)	1.609	kilometer per hour (km/h)

* This is an edited, reformatted and augmented version of U.S. Geological Survey Circular 1347, dated 2010.

Introduction

Of all the potential threats posed by climatic variability and change, those associated with water resources are arguably the most consequential for both society and the environment (Waggoner, 1990). Climatic effects on agriculture, aquatic ecosystems, energy, and industry are strongly influenced by climatic effects on water. Thus, understanding changes in the distribution, quantity and quality of, and demand for water in response to climate variability and change is essential to planning for and adapting to future climatic conditions. A central role of the U.S. Geological Survey (USGS) with respect to climate is to document environmental changes currently underway and to develop improved capabilities to predict future changes. Indeed, a centerpiece of the USGS role is a new Climate Effects Network of monitoring sites. Measuring the climatic effects on water is an essential component of such a network (along with corresponding effects on terrestrial ecosystems).

The USGS needs to be unambiguous in communicating with its customers and stakeholders, and with officials at the Department of the Interior, that although modeling future impacts of climate change is important, there is no more critical role for the USGS in climate change science than that of measuring and describing the changes that are currently underway. One of the best statements of that mission comes from a short paper by Ralph Keeling (2008) that describes the inspiration and the challenges faced by David Keeling in operating the all-important Mauna Loa Observatory over a period of more than four decades. Ralph Keeling stated: "The only way to figure out what is happening to our planet is to measure it, and this means tracking changes decade after decade and poring over the records."

There are three key ideas that are important to the USGS in the above-mentioned sentence. First, to understand what is happening requires measurement. While models are a tool for learning and testing our understanding, they are not a substitute for observations. The second key idea is that measurement needs to be done over a period of many decades. When viewing hydrologic records over time scales of a few years to a few decades, trends commonly appear. However, when viewed in the context of many decades to centuries, these short-term trends are recognized as being part of much longer term oscillations. Thus, while we might want to initiate monitoring of important aspects of our natural resources, the data that will prove to be most useful in the next few years are those records that already have long-term continuity. USGS streamflow and groundwater level data are excellent examples of such longterm records. These measured data span many decades, follow standard protocols for collection and quality assurance, and are stored in a database that provides access to the full period of record.

The third point from the Keeling quote relates to the notion of "poring over the records." Important trends will not generally jump off the computer screen at us. Thoughtful analyses are required to get past a number of important but confounding influences in the record, such as the role of seasonal variation, changes in water management, or influences of quasi-periodic phenomena, such as El Niño-Southern Oscillation (ENSO) or the Pacific Decadal Oscillation (PDO). No organization is better situated to pore over the records than the USGS because USGS scientists know the data, quality-assure the data, understand the factors that influence the data, and have the ancillary information on the watersheds within which the data are collected.

To fulfill the USGS role in understanding climatic variability and change, we need to continually improve and strengthen two of our key capabilities: (1) preserving continuity of long-term water data collection and (2) analyzing and interpreting water data to determine how the Nation's water resources are changing.

Understanding change in water resources to date and predicting change into the future must be done in full recognition of the other factors that influence water availability, including changes in water use, land use, the design and operation of water infrastructure, and the depletion of groundwater. There is widespread debate about the relative importance of non-climatic factors versus climatic factors in determining water conditions and characteristics over the coming decades (Lins and Stakhiv, 1998). Differentiating climatic from nonclimatic effects is, therefore, a critical component of any effective assessment of how climate has, is, and will affect the Nation's water resources. Moreover, such assessments are critically dependent upon a quality-assured, spatially and temporally comprehensive water-resources database.

For more than 100 years, the USGS has been monitoring the Nation's rivers and groundwater resources, and currently maintains the world's most sophisticated information system of surface-water, groundwater, water-quality, and water-use data. The USGS is uniquely positioned to play a central role in providing both data and analyses to address the recently recognized increment of hydrologic uncertainty associated with climatic variability and change (Moss and Lins, 1989).

The USGS draws upon two decades of climate change research related to hydroclimatology—the study of hydrologic events and conditions within their climatologic context. USGS research provides an empirical basis for integrating the physical sources of variability in a hydrologic time series with the statistical properties of the varying driving force itself. This research also highlights several themes where our incomplete understanding of important hydrologic processes and conditions continues to serve as an impediment to the broader incorporation of climate change information in water-resources planning and design. Thus, expanding our understanding of these processes is key to developing improved capabilities for dealing with climate uncertainty in the field of water-resources planning and management.

UNDERSTANDING CLIMATE AND WATER

Climate affects all aspects of hydrology and water resources through the water budget. Among the more critical elements of this issue to the Nation are how climate is affecting low flows and drought, high flows and flooding, and the seasonality (timing) of streamflow. Each of these factors can only be meaningfully understood where there exists a spatially and temporally adequate record of high-quality hydrologic observations, such as those derived from the USGS streamgaging network.

Climate-Sensitive Hydrologic Data

Information on the flow of rivers is a national asset that safeguards lives and property and ensures adequate water resources for a healthy economy (Hirsch and Norris, 2001). As noted by the National Research Council (1999), *"Stream-flow data collected by the USGS ... form the cornerstone for national, regional, and local efforts ... by providing continued, up-to-date information about water conditions and understanding of hydrologic phenomena."* In this context, an important component of the USGS National Streamflow Information Program (NSIP) is to provide data that are minimally affected by human activities for the study of regional streamflow characteristics and assessments of trends due to factors such as changes in climate, land use, and water use. Within NSIP, the stations providing such data are referred to as sentinel watersheds, and most are derived from the existing USGS Hydro-Climatic Data Network (HCDN). Sentinel watersheds also include the 36 discharge-monitoring stations of the USGS Hydrologic Benchmark Network (HBN).

HCDN stations were identified to provide a unique database of high-quality, climate-sensitive observations, specifically for evaluating climatic effects on streamflow (Slack and Landwehr, 1992). For nearly two decades, the observations from these stations have been the principal data used by researchers worldwide in studies of United States hydro-climatic patterns, variations, and trends. The criteria used to identify these stations have been adopted by hydrological services in other countries and, more recently, by the World Meteorological Organization's Commission for Hydrology as part of its efforts to establish a global climate-sensitive stream-flow database [World Meteorological Organization (WMO), 2004]. When established in 1992, the HCDN consisted of 1,659 stations. Over time, many of these stations have been discontinued and there are numerous gaps in the geographical distribution of stations, particularly in the western United States and at higher elevations. An effort is currently underway to update the listing of HCDN stations using a predictive model of climate-sensitive watershed conditions based on hydrological alteration criteria (Carlisle and others, 2009).

The Hydrologic Benchmark Network (HBN) was established in 1963 to provide long-term measurements of streamflow and water quality in areas that are minimally affected by human activities. These data are used to study long-term trends in surface-water flow and water chemistry and as a benchmark against which to identify changes in flow and chemistry in developed watersheds (Murdoch and others, 2005). HBN, HCDN, and the corresponding NSIP sentinel watersheds are, therefore, logical and essential elements of the incipient USGS Climate Effects Network.

Climate-Directed Hydrologic Analyses

The effect of climate (particularly precipitation and temperature) on hydrology (particularly streamflow) has been a focus of research among USGS hydrologists for six decades. The classic hydroclimatic study by Langbein and others (Geological Survey Circular 52, 1949) graphically related mean annual runoff in the United States to mean annual precipitation and temperature, and was a standard reference used in water balance work through the 1980s (fig. 1). Some of this reference's more enduring aspects were its illustration of (1) how runoff, for a given annual precipitation, decreases as temperature increases; (2) how runoff, for a given temperature, increases with precipitation; and (3) how the numerical difference between precipitation and runoff, for a given temperature, increases with precipitation and ultimately reaches a constant representing the limiting rate of evapotranspiration. This work served as the basis for many early climate change, water-resources impact studies during the 1980s and early 1990s, when the emphasis shifted to studying such interactions using numerical models.

During the past 15 years, studies of climate change impacts on water have increasingly made use of general circulation model (GCM) simulations of future climate states [Intergovernmental Panel on Climate Change (IPCC), 1995, 2001, 2007]. Most such studies use model-simulated temperature and precipitation to draw inferences about water-related impacts. As most projected climate change effects related to water resources (for example, changes in flooding, drought, and the seasonality of snowmelt-generated streamflow) are subgrid (local) scale processes, they are not resolved explicitly within the models. Rather, they are inferred from the changes in temperature and precipitation simulated for the grid cell surrounding the location of interest.

The temperature increases simulated by the models occur in response to rising concentrations of atmospheric carbon dioxide (CO_2) that, in turn, cause increases in both evaporation and precipitation. This increase in atmospheric heat and moisture is frequently referred to as indicating an "enhanced hydrological cycle." Since being reported initially in the Second Assessment Report (Intergovernmental Panel on Climate Change (IPCC), 1995), the enhanced hydrological cycle has been cited generally as the explanation for why floods and droughts are expected to increase with increasing CO_2 concentrations. Changes in the timing of snowmelt-generated streamflow are mentioned as resulting from increasing winter and spring temperatures, whereby mountain snows melt and resultant peak streamflows occur earlier in the year. Increasing spring and summer season temperatures lengthen the warm season and the amount of evaporation, and are expected to reduce further the water supply in lakes, reservoirs, and rivers by summer's end.

Are these water-related impacts of CO_2-induced climate change already manifesting themselves within the United States? Answering this question is much more complicated than it may appear. Not all changes occurring in the climate system are the result of increased CO_2 concentrations. Indeed, the extent to which observed climatic variations over the past three decades are the result of CO_2 emissions versus changes in natural factors (for example, El Niño-Southern Oscillation (ENSO), Pacific Decadal Oscillation (PDO), Atlantic Multidecadal Oscillation (AMO), solar variations, etc.) is unknown. However, existing streamflow data provide the requisite empirical basis for understanding both short-term (seasonal to decadal) hydrologic variations and long-term (multidecadal to centennial) hydrologic trends, and for testing hypotheses regarding the hydrologic response to climate change. These same data also make it possible to differentiate hydrologic changes due to climate from those resulting from other important forcing factors, such as urbanization, groundwater development, dam building, changes in the area of irrigated agriculture, changes in the area of row-crop agriculture, and changes in water withdrawals or consumption.

The following four topical segments represent issues of critical concern to the Nation's water-resources planning and management communities, and have been highlighted as possibly being adversely impacted by simulated changes in climate.

The USGS has conducted a number of studies on these issues during the past two decades (many of which are cited), but never in the context of a systematic or comprehensive program of investigations. As a result, substantive and crucial gaps remain in each issue area, particularly with respect to understanding operative climate forcing. Collectively, the contents of these topical segments constitute a unique program of hypothesis testing on key assumptions associated with climate change impacts on water resources. Critical questions and issues that the USGS proposes to address are highlighted in bold print.

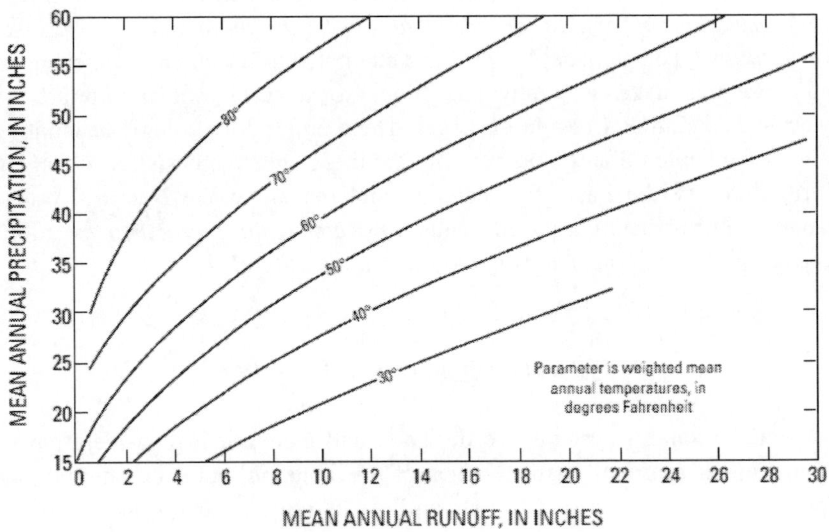

Source: Langbein and others, 1949.

Figure 1. Relation of annual runoff to precipitation and temperature.

WATER BUDGET

Over time, changes in mean discharge can pose substantial problems, but can also pose opportunities for water-resources managers. Although the effect of climatic change on hydrologic extremes has tended to receive the most attention, Lins and Cohn (2003) analyzed the sensitivity of streamflow to changes in precipitation and found that the precipitation sensitivity of mean streamflow is much greater than that of flood flows. They also found that the greater the return period of the flood, the lower the precipitation sensitivity. This finding suggests that changes in climate are more likely to produce changes in the mean state of hydrological regimes than in hydrological extremes. From a water-resources planning and management perspective, therefore, the more crucial questions to answer may be associated with how to take advantage of opportunities posed by potential increases in mean discharge, as well as how to prepare appropriately for the threats posed by potential decreases in mean discharge.

The following questions identify the types of information that USGS researchers need to provide to be of maximum value to water managers: *Have multidecadal changes in mean discharge occurred and, if so, are these variations distinct from observed centennial changes? Have changes occurred in the relation of annual or seasonal discharge to annual or seasonal precipitation?* To date, few studies have attempted to document such changes, and those that have, generally have done so at scales not appropriate for resource planning and design purposes. There are 352 hydrologic accounting units in the United States, and not only are these units the basis for managing NSIP, they are the basis for planning and managing the Nation's water-resources infrastructure as well. Thus, *answers to the above two questions need to be sought at the hydrologic-accounting unit level.*

SEASONAL STREAMFLOW

Water-resources systems are designed and managed in consideration of the prevailing seasonal characteristics of precipitation, both the amount and type, and their effect on seasonal streamflow. In most water-resources regions of the United States, streamflow decreases to its minimum value in September, coincident with the end of the water year.[1] Exceptions are the Souris-Red-Rainy and Upper Colorado water-resource regions where snow and river ice in mid-winter cause the annual minimum to occur in February (Lins and Slack, 2005). Maximum streamflow is more temporally variable, occurring over the 5-month period from February (California water-resource region) to June (Missouri water-resource region), and generally reflects the timing of snowmelt and (or) the month of maximum rainfall.

Concern has been raised that in a scenario where the climate is warmer than today's climate, reservoir contents in regions where streamflows are primarily snowmelt driven, rather than rainfall driven, could be reduced earlier in the summer than at present, leading to reduced water availability during the season of peak water demand. This reduction in availability would result from a combination of increased evaporation and transpiration from warmer temperatures and a lengthening of the warm season, as well as increased irrigation demand. Moreover, several recent studies have shown that in mountainous areas of the Western United States and in northern New England, the midpoint in the total volume of streamflow for the first 6 months of the calendar year appears to be occurring 1 to 3 weeks earlier than in the past (Dettinger and Cayan, 1995; McCabe and Clark, 2005; Stewart and others, 2005; Hodgkins and Dudley, 2006); however, no reduction in summer season flows has been observed (Lins and Slack, 2005).

Tracking and understanding the reasons for such shifts are important because of the impact they can have on water-resources and aquatic ecosystems management. For example, reservoir rule curves, which were developed using streamflow observations from decades ago, may need to be modified to ensure the maintenance of adequate municipal and agricultural water supplies, in-stream flow for aquatic ecosystems, or to provide flood protection.

Despite the insights gained from recent USGS research, numerous questions remain. *Are the shifts that are currently taking place important to water management (water supply, flood mitigation, and aquatic ecosystem integrity)? If not, why not, and how long would a shift have to continue before it becomes important to these kinds of issues?* Although climate change has been implicated as the likely cause for the observed hydrologic changes, *to what extent could multidecadal climatic-oceanic oscillations or changes in groundwater development or water use be responsible for these observed changes?* Being able to differentiate and quantify the water-resources response to both climatic and nonclimatic drivers of hydrologic change is critical for any effective program of adaptive water management.

Low-Flow Characteristics

The characteristics of low flows in rivers and streams are important metrics for water managers who must meet a growing number of water-supply requirements, particularly when low flows persist over an extended period as during drought. Low flows are also critical for managing water quality where pollutant concentrations must be maintained below regulatory thresholds required by the Clean Water Act. Low flows, which regularly occur during late summer in most of the United States, can be affected by changes in precipitation, evapotranspiration, and, in snow-melt dominated areas, by shifts in seasonality (that is, late winter and spring temperatures). Annual low flows are distinct from drought flows primarily in terms of their temporal duration and magnitude. The IPCC (1995, 2001, 2007) model simulations indicate drying in midcontinental areas during the summer and the possibility of more severe droughts. However, empirical studies of trends in low flows have not shown evidence of a general decrease in low flows (Svensson and others, 2005). Indeed, several studies have found substantial increases in low flows in the United States over the past half century (Lins and Slack, 1999; 2005; McCabe and Wolock, 2002; fig. 2).

Explaining variations and changes in low streamflow is complicated by the fact that low flows are sensitive to nonclimatic factors, particularly to manmade water storage, water use, and groundwater depletion. The following research questions are critical to gaining additional insight into how and to what extent climate is affecting low flows: *Is there a pattern of change in 1-day, 7-day, 14-day, or 28-day low flows? Are we seeing an unexpected number of new record low flows (more than expected or less than expected by chance)? If low flows are changing, what are the relative influences of changing water use, groundwater depletion, reservoir storage, or intrabasin transfers on low flows? To what extent is it possible to predict regional low flows seasonally and annually based on North Atlantic and North Pacific sea-surface temperatures as hypothesized by McCabe and others (2004)?*

Flood and High-Flow Characteristics

Floods are the most costly natural hazard in the United States, causing an average of 140 fatalities and $5 billion in damages each year (O'Connor and Costa, 2003). Any change in the factors that affect flood hazards can, therefore, dramatically impact the Nation's social and economic well-being.

Increased urban and suburban development on flood plains has, for example, substantially increased the mortality and economic losses from floods (Pielke and Downton, 2000). Moreover, as noted earlier, climate model simulations have been interpreted as indicating that more and larger floods will occur in response to a warming-induced, enhanced hydrological cycle. To date, numerous studies of annual maximum (daily mean) streamflow have found no evidence of an increasing trend in high streamflow (Lins and Slack, 1999; 2005; McCabe and Wolock, 2002; Douglas and others, 2000; and Kundzewicz and others, 2005).

One study, using different data (annual maximum based on monthly mean stream-flow) and methods, reported an increase in the frequency of floods having an exceedance probability <0.01 during the 20th century (Milly and others, 2002). A concern associated with all of these studies, however, is that annual maximum streamflow records include many observations that are not true (out-of-bank) floods. A more representative dataset for assessing flood trends is the USGS peak flow file, maintained within the USGS National Water Information System (NWIS) (*http://nwis. waterdata.usgs.gov/ usa/nwis/peak*). Peak flow data represent the maximum instantaneous discharge of water passing a streamgage during each year; although it should be noted that in any given year, even these values may not necessarily represent a true (out-of-bank) flood as most streams do not experience flooding every year.

An example of a peak flow record is shown in figure 3 for the USGS streamgage on the Red River of the North at Grand Forks, North Dakota, for the period 1882–2009. The history of flooding at this site and others in the region is complex. The graph demonstrates the importance of having long-term flow records and carefully and thoroughly examining those records. For example, an analysis of this record for the period 1930–2009 would conclude that there is a substantial upward trend in flood discharges that appears to continue to the present. An association between this trend and contemporaneous increases in greenhouse gas concentrations or land-use change would seem highly plausible if only this part of the record is considered.

However, if the entire period of record is assessed, the conclusion is quite different. Considering the entire period, a more likely conclusion is that floods arise from at least two substantially different statistical populations. One population has very large annual peak discharges that prevailed in the years from 1882 to about 1900 and again after about 1942; the other population has much smaller peaks that occurred from about 1900 to 1942.

Thus, in considering the entire period of record, it becomes apparent that very large floods were common well before the onset of large increases in greenhouse gases or extensive land modifications. In this case, the data strongly support persistence in climate states as the key mechanism for increased peak flows, rather than a human forcing of change.

One of the most important contributions that the USGS can make to understand better climatic variability and change is to explore the numerous streamflow records that have been collected over the past century, and in some instances longer, and examine flood flows, average flows, seasonal flows, and low flows.

Critically, this exploration needs to include investigations of both the patterns of change and their relation to a wide range of potential driving variables, including such atmospheric, oceanic, and terrestrial features such as El Niño and the Pacific Decadal Oscillation, as well as greenhouse gas concentrations, and land-use changes in the basin. Water-resources managers and those responsible for flood-plain management and land-use decisions, are looking for answers to how streamflow is changing and where it may be headed in the future.

Analysis of datasets, such as the example of annual peak flows from the Red River of the North, is of paramount importance in support of understanding the implications of climatic variability and change for water resources and water-related hazards.

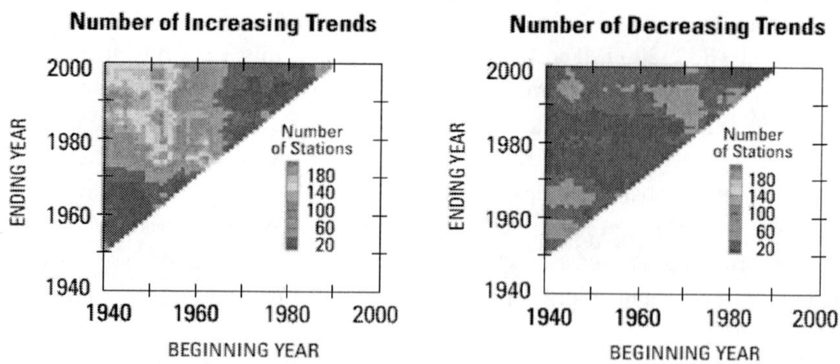

Source: McCabe and Wolock, 2002.

Figure 2. Number of streamgaging stations with statistically significant ($p \leq 0.05$) increasing and decreasing trends in annual minimum daily streamflow for various periods at least 10 years in length during 1941 through 1999.

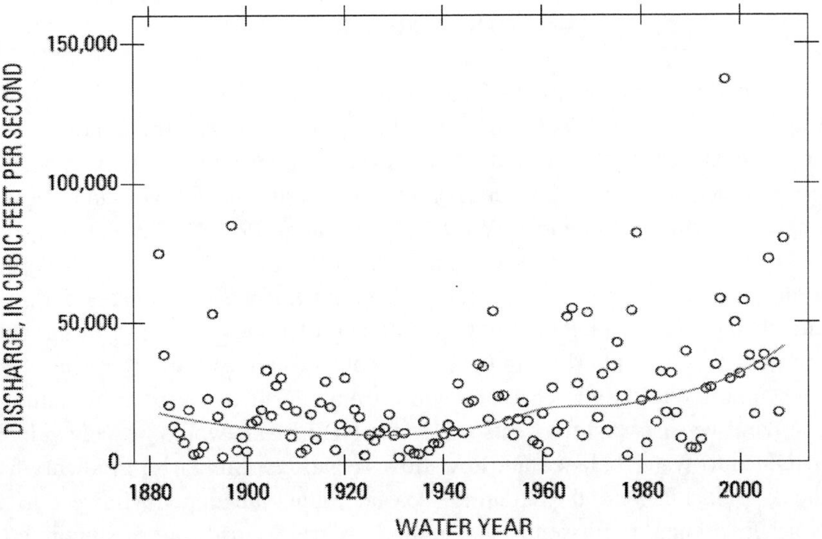

Figure 3. Annual peak discharges, at the USGS streamgage, Red River of the North, Grand Forks, North Dakota, 1882 through 2009. Figure also includes (in blue) a locally weighted smoothed estimate of the changes in the mean flood magnitude over the period of record.

Recently, the USGS implemented procedures for assessing the quality of its peak flow records, and for identifying and correcting errors in the records. An appropriate flood database now exists that can be used to help discern if a pattern of change in observed annual peak discharges has occurred. In particular, the questions of greatest interest include: Is there a pattern of change in annual flood peak discharges or flood volumes? Is there a change in the timing of annual peak discharges?

Is there a change in the number of largest floods (as defined by exceedances of a particularly high threshold)? Are there an unexpected number of new record peaks (either more or less than would be expected by chance)? Can the role that dams play in changing high-flow characteristics be quantified? Can any observed changes in peak flows be related unambiguously to changes in climatic factors and (or) land-use factors?

Geographical Design

Obtaining answers to the questions highlighted in the previous section requires a mix of national and regional investigations. Hydroclimatic patterns and events across the United States are a response to highly varying, but geographically persistent, air-mass boundary conditions. Of particular regional interest is the mountainous West, the Northern Plains (Dakotas), and the Northeast, where snowpacks and seasonal river ice create enhanced hydroclimatic sensitivities that can presage critical water-resources variations and change. The gradient from the Continental Divide to just east of the 100th meridian is a highly engineered water-resources environment, with substantially different characteristics from north to south. A suite of comparative east-west transects across the High Plains would provide valuable insight into water balance and low-flow variations. Similarly, gradients from the Appalachians to the Atlantic Coastal Plain, the upper Midwest to the Northern Plains, north-south across the semiarid to arid Intermountain West, and the hurricane-prone areas from Texas to New England each have unique climate-driven water-resources issues that vary considerably from year to year, and where documentation of longer term variations and changes is critical to ensuring the reliability of the Nation's water supply.

In addition to geographically focused studies, there also is a need for issue-oriented investigations that can capitalize on unique characteristics of particular river basins. Issues associated with basins having a large amount of storage in relation to flow, as in the Colorado River Basin; a high percentage of urbanization, such as in the Delaware River Basin; or a high percentage of agriculture or irrigated agriculture, such as in the Missouri River Basin, all pose situations that can be difficult to analyze yet be of enormous consequence to the Nation's economy and resource security. In this regard, studies that focus on reference sites, such as the incipient USGS Climate Effects Network and the USGS Hydrologic Benchmark Network, are vital to identifying and differentiating climatically driven variations and trends from human-induced landscape effects.

In summary, the USGS is uniquely poised to make a major contribution to the understanding of the role that climatic variability and change can have on our Nation's water resources and aquatic ecosystems. The key attributes that will enable the USGS to make that contribution are: (1) the ability to ensure the continued monitoring of a wide range of hydrologic systems (particularly those that already have long records and limited influences of man on the landscape); and (2) the capacity to substantially increase efforts to analyze existing records to understand changes that are taking place and to differentiate the relations of water resources to climate variation, climate change, and direct human activities on the landscape. "The only way to figure out what is happening to the planet is to measure it, and this means tracking changes decade after decade and poring over the records" (Keeling, 2008). This statement should be the guiding principle by which the USGS Climate Effects Network is implemented.

APPENDIX 1. LIST OF ABBREVIATIONS

AMO	Atlantic Multidecadal Oscillation
CO2	carbon dioxide
ENSO	El Niño-Southern Oscillation
GCM	general circulation model
HBN	USGS Hydrologic Benchmark Network
HCDN	USGS Hydro-Climatic Data Network
IPCC	Intergovernmental Panel on Climate Change
NSIP	USGS National Streamflow Information Program
NWIS	USGS National Water Information System
PDO	Pacific Decadal Oscillation
USGS	U.S. Geological Survey
WMO	World Meteorological Organization

REFERENCES CITED

Carlisle, D.M., Falcone, James, Wolock, D.M., Meador, M.R., and Norris R.H., 2009, Predicting the natural flow regime: models for assessing hydrological alteration in streams: *River Research and Applications,* v. 26, no. 2, p. 118–136. (Also available at http://dx.doi.org/10.1002/rra.1247.)

Dettinger, M. D., and Cayan, D. R., 1995, Large-scale atmospheric forcing of recent trends toward early snowmelt runoff in California: *Journal of Climate*, v. 8, p. 606–623.

Douglas, E. M., Vogel, R. M., and Kroll, C. N., 2000, Trends in floods and low flows in the United States: impact of spatial correlation: *Journal of Hydrology*, v. 240, p. 90–105.

Hirsch, R.M., and Norris, J.M., 2001, National streamflow information program: implementation plan and progress report: USGS Fact Sheet FS–

048–01, 6 p. (Also available at http://pubs.er.usgs.gov/usgspubs/ fs/ fs04801.)

Hodgkins, G.A., and Dudley, R.W., 2006, Changes in the timing of winter-spring streamflows in eastern North America, 1913-2002: Geophysical Research Letters, v. 33, L06402. (Also available at *http://dx.doi. org/10.1029/2005GL025593.)*

Intergovernmental Panel on Climate Change (IPCC), 1995, Climate Change 1995: The Science of Climate Change, Contribution of Working Group I to the Second Assessment Report of the Intergovernmental Panel on Climate Change: Cambridge University Press, Cambridge, UK.

Intergovernmental Panel on Climate Change (IPCC), 2001, Climate Change 2001: The Scientific Basis. Contribution of Working Group I to the Third Assessment Report of the Intergovernmental Panel on Climate Change: Cambridge University Press, Cambridge, UK.

Intergovernmental Panel on Climate Change (IPCC), 2007, Climate Change 2007: The Physical Science Basis, Contribution of Working Group I to the Fourth Assessment Report of the Intergovernmental Panel on Climate Change: Cambridge University Press, Cambridge, UK.

Keeling, R., 2008, Recording the Earth's vital signs: Science, v. 319, p. 1771–1772.

Kundzewicz, Z.W., Graczyk, D.J., Maurer, T., Przymusinska, I., Radziejewski, M., Svensson, C., and Szwed, M., 2005, Trend detection in river flow series: 1. Annual maximum flow: *Hydrological Sciences Journal*, v. 50, p. 797–810.

Langbein, W.B., and others, 1949, Annual runoff in the United States: U.S. Geological Survey Circular 52, 14 p. (Also available at *http://pubs.er. usgs.gov/usgspubs/cir/cir52.)*

Lins, H.F., and Cohn, T.A., 2003, Floods in the greenhouse: spinning the right tale *in* Thorndycraft, V.R., Benito, G., Barriendos, M., and Llasat, M.C., eds., Palaeofloods, Historical Data and Climatic Variability: Applications in Flood Risk Assessment, Centro de Ciencias Medioambientales, Madrid, p. 263–268.

Lins, H.F., and Slack, J.R., 1999, Streamflow trends in the United States: *Geophysical Research Letters*, v. 26, p. 227–230.

Lins, H.F., and Slack, J.R., 2005, Seasonal and regional characteristics of U.S. streamflow trends in the United States from 1940–1999: *Physical Geography*, v. 26, p. 489–501.

Lins, H.F., and Stakhiv, E.Z., 1998, Managing the nation's water in a changing climate: *Journal of the American Water Resources Association*, v. 34, p. 1255–1264.

McCabe, G.J., and Clark, M.P., 2005, Trends and variability in snowmelt runoff in the western United States: *Journal of Hydrometeorology*, v. 6, p. 476–482.

McCabe, G.J., and Wolock D.M., 2002, A step increase in streamflow in the conterminous United States: *Geophysical Research Letters*, v. 29, p. 2185–2188.

McCabe, G.J., Palecki, M.A., and Betancourt, J.L., 2004, Pacific and Atlantic Ocean influences on multidecadal drought frequency in the United States: Proceedings, *National Academy of Sciences*, v. 101, p. 4136–4141.

Milly, P.C.D., Wetherald, R.T., Dunne, K.A., and Delworth, T.L., 2002, Increasing risk of great floods in a changing climate: *Nature*, v. 415, p. 514–517.

Moss, M.E., and Lins, H.F., 1989, Water resources in the twenty-first century—a study of the implications of climate uncertainty: U.S. Geological Survey Circular 1030, 25 p. (Also available at *http://pubs.er.usgs.gov/usgspubs/cir/ cir1030*.)

Murdoch, P.S., McHale, M.R., Mast, M.A., and Clow, D.W., 2005, The U.S. Geological Survey hydrologic benchmark network: U.S. Geological Survey Fact Sheet 2005–3135, 6 p. (Also available at *http://ny.water. usgs.gov/pubs/fs/ fs20053135/*.)

National Research Council, 1999, Hydrologic Hazards Science at the U.S. Geological Survey: National Academy Press, Washington, D.C., 79 p.

O'Connor, J.E., and Costa, J.E., 2003, Large floods in the United States: where they happen and why: *U.S. Geological Survey Circular* 1245, 13 p. (Also available at *http://pubs. er.usgs.gov/usgspubs/cir/cir1245.)*

Pielke, R.A., and Downton, M.W., 2000, Precipitation and damaging floods: trends in the United States, 1932–97: *Journal of Climate*, v. 13, p. 3625–3637.

Slack, J.R., and Landwehr, J.M., 1992, Hydro-climatic data network: a U.S. Geological Survey streamflow data set for the United States for the study of climate variations, 1874–1988: U.S. Geological Survey Open-File Report 92–129, 193 p. (Also available at *http://pubs.usgs.gov/ of/1992/ofr92-129/*.)

Stewart, I.T., Cayan, D.R., and Dettinger, M.D., 2005, Changes toward earlier streamflow timing across western North America: *Journal of Climate*, v. 18, p. 1136–1155.

Svensson, C., Kundzewicz, Z.W., and Maurer, T., 2005, Trend detection in river flow series: 2. Flood and low flow index series: *Hydrological Sciences Journal*, v. 50, p. 811–824.

Waggoner, P.E., ed., 1990, Climate Change and U.S. Water Resources: Report of the American Association for the Advancement of Science Panel on Climatic Variability, Climate Change and the Planning and Management of U.S. Water Resources: John Wiley and Sons, New York, 496 p.

WMO (World Meteorological Organization), 2004, Commission for Hydrology, 12th Session: *Abridged Final Report with Resolutions and Recommendations*: WMO-No. 979, Geneva, 55 p.

End Note

[1] The 12-month period from October through September. The water year is designated by the calendar year in which it ends and which includes 9 of the 12 months. For example, the year ending September 30, 1992, is called the "1992 water year."

In: Climate Change
Editor: Aileen Stablum

ISBN: 978-1-63117-584-8

Chapter 3

REPORT TO CONGRESS: STRENGTHENING THE SCIENTIFIC UNDERSTANDING OF CLIMATE CHANGE IMPACTS ON FRESHWATER RESOURCES OF THE UNITED STATES*

Federal Interagency Panel on Climate Change and Water Data and Information

EXECUTIVE SUMMARY

This report responds to the requirements of Section 9506 of the Omnibus Public Lands Act [Public Law (PL) 111–11; Appendix A] calling for a report to Congress that describes the impacts of global climate change on freshwater resources of the United States, and identifies key actions to improve the Nation's capacity to detect and predict changes in freshwater resources that are likely to result from a changing climate. The steps described in the report are intended to help decision-makers and water resource managers by facilitating improvements in observational data, data acquisition, and modeling capabilities.

---

* This is an edited, reformatted and augmented version of a report dated August 2011.

Freshwater resources are vulnerable to a number of stressors, and a changing climate can exacerbate other stressors and creates new risks to the Nation's food security, energy security, and overall economic health. Some key impacts of climate change on water resources that have been identified to-date (U.S. Global Change Research Program, 2011) include:

- Increasing air and water temperatures;
- Declining rainfall amounts in some areas and increasing amounts in other areas, and decreasing proportion of precipitation that falls as snow;
- Changes in the timing of snowmelt runoff;
- Decline in the mass of water stored in glaciers;
- More intense rainfall and storm events;
- Rising sea levels; and
- Changing quality of freshwater, coastal, and ocean waters.

The Nation invests considerable resources in monitoring, mapping, evaluating, assessing, modeling, and managing water resources. Many of the existing observational water data networks, models, and hydro-statistical methods were developed for specific users and pre-date recent advances in climate change science. As a result, these systems (networks, methods, and models) were not designed to consider climate-induced stressors, to account for non-stationary hydroclimatic processes, or to evaluate the effectiveness of climate change mitigation and adaptation strategies. Today, there is a need and an opportunity to modernize hydroclimatic data networks and climate-relevant data collection, data management, mapping, modeling, and information dissemination. Of particular importance is maintenance and strengthening of long-term *in situ* and remote observational capabilities to detect change.

This report provides a general overview of the challenges that a changing climate poses for water resource managers in the context of other water-resources stressors. In particular, the report considers water resources measurement and modeling systems that are relevant to climate change adaptation, as required by Section 9506. Recommendations are focused on Federal actions to strengthen these systems to inform water management decisions at the Federal, State, and local levels. This report draws from and builds on a number of recent climate and water documents that have been produced across the Federal, State, local, Tribal, and private water sectors.

Observational data, the networks that provide these data, the systems that make these data available, and the models that enable projections of future climate conditions are critical for decision-making. Hydrologic data, science, and technology need to be coordinated and improved to promote the ready interoperability of data and models, and ultimately provide decision-makers with more effective information, tools, and services.

Findings

The key findings for each of the six "review elements" included in the statute (that is, Section 9506 of PL 111-11) are summarized below; findings for review elements 1 and 2 are combined.

Review Elements 1 and 2. Assess Adequacy of Current Hydro-Climatological Observation Networks and Identify Gaps in Current Networks

Finding 1: Broad Support for Improved Monitoring Systems. There is broad and continuing support from a wide constituency to sustain and strengthen the Nation's hydroclimatic monitoring systems, including research to support those systems. In particular, adaptive management and water-resources planning tools are only effective when supported by strong data networks.

Finding 2: Streamflow Data. The National Streamflow Information Program was begun in 2003 with 5 program goals. These goals are to (1) develop a stable stream-gaging network; (2) conduct intensive data collection during floods and droughts; (3) conduct regional and national streamflow assessments; (4) improve data delivery; and (5) invest in methods development research. Current progress toward meeting the goals include 15 percent of the stable stream-gaging network in place with some improvements in data delivery.

Finding 3: Streamflow Information Coordination. Streamflow information is collected by a variety of Federal and State agencies, generally for agency-specific purposes. Tremendous benefits could be gained for the Nation if these disparate data sets were more broadly available, interoperable, and included information on methods of collection, quality, and accuracy.

Finding 4: Reservoir and Lake Data. Reservoir water-level and volume information is collected by a variety of Federal agencies and private entities. Data are not widely available, and reservoir volume information generally is out of date. No data are available for many natural lakes, and a national lake level network does not exist.

Finding 5: Groundwater-Level Monitoring. The Advisory Committee on Water Information's Subcommittee on Groundwater has developed a comprehensive national groundwater monitoring network that could provide data and information necessary for planning, management, and development of the Nation's groundwater resources in a sustainable manner. This network was authorized in Section 9507 of PL 111-11.

Finding 6: Soil Moisture. Numerous specialized soil moisture networks exist across the Nation, but there is no design for a national soil moisture network. The NASA Soil Moisture Active Passive Mission, planned for 2014, in conjunction with a robust *in situ* network for calibration, could provide adequate soil moisture data for many applications at regular intervals over much of the country if these efforts were implemented in a coordinated manner.

Finding 7: Evapotranspiration. Operational evapotranspiration data can be obtained from remotely-sensed information. *In situ* measurements of energy, carbon, and water fluxes above the canopy provide valuable information on water budgets (as well as energy and carbon budgets) and are valuable for improving remotely-sensed evapotranspiration estimates, but the number of such installations is relatively small.

Finding 8: Precipitation Frequency-Duration-Intensity Estimates. A systematic and consistent approach that includes recent data and improved statistical techniques is needed for updating precipitation intensity-duration-frequency estimates.

Finding 9: *In Situ* Precipitation Measurement. There is an opportunity to form a robust precipitation measurement network in the U.S. with existing gages operated by numerous Federal, State, and local agencies, research organizations, and private citizens. However, lack of coordination, protocol, and communication among the organizations impedes formation of such a network. Because of this lack of coordination among precipitation gage operators, there has been no broad analysis of precipitation measurement needs.

Finding 10: Remotely-Sensed Precipitation Measurements. Radar and satellite precipitation measurement capabilities have provided precipitation estimates over areas not covered by ground-based gages. These estimates, however, rely on gage data for calibration and bias correction.

Finding 11: Snowpack Monitoring. The Natural Resources Conservation Service's SNOTEL (SNOwpack TELemetry) network operates in 12 western States and Alaska. NRCS has proposed to coordinate and provide technical leadership for collection and dissemination of snow information over a greater area of the U.S., which could enhance products from NOAA's Snow Data Assimilation System.

Finding 12: Glacier Monitoring. Glacier monitoring is important, given the potentially large effect glacier meltwater could have on sea level and ocean circulation. A combination of ground-based and satellite monitoring is warranted. As in many monitoring efforts, there is some lack of coordination among existing monitoring activities.

Finding 13: Water-Quality Networks. Extensive water-quality monitoring is conducted in the U.S., and various networks have evolved to meet user needs. Integration of data from the various networks could offer benefits, if data were comparable and well integrated. The proposed National Monitoring Network offers a sound basis for understanding fluxes of key contaminants from major basins across the Nation.

Finding 14: Waterborne Pathogens. Prevalence of waterborne pathogens, including harmful algal blooms, likely will change in response to rising temperatures, as well as to alterations in flow characteristics and contaminant loadings. Systems for tracking these changes are not well developed, nor are they well integrated with geospatial, water-quality, and public health information.

Finding 15: Wetlands. Freshwater and coastal wetlands are sensitive to changes in hydrologic and sea-level conditions. The National Wetland Assessment should provide valuable information on the conditions of a sampling of the Nation's wetlands, and a basis for future comparisons, although information on specific systems will not be available. Implementation of the new wetlands mapping standard needs to be expedited.

Finding 16: Withdrawals and Consumptive Use of Water. Water withdrawal and consumptive use data, particularly as related to the Nation's energy portfolio, are inadequate for making long-term decisions about water availability in the U.S.

Finding 17: Environmental Flows. New data and research on the environmental flow requirements, including pre- and post-alteration information, are needed for the restoration and maintenance of healthy aquatic ecosystems.

Review Element 3. Improve Data Management to Increase Utility of the Information That Is Collected and Efficiency of Data Acquisition and Reporting

Finding 18: Interoperable Data Systems. Ready access to the full range of hydroclimatic data collected by government agencies and other interests is inadequate. Data are collected using a range of protocols, which are not always documented, and are archived in a variety of ways, from modern relational databases to paper copies in files. There is much to be gained from use of consistent documentation standards and improvements in interoperability of data systems.

Finding 19: Data and Decision-Making. In general, hydroclimatic data are insufficiently integrated (or readily integratable by the user community) to support important management decisions, and hydroclimatic data are inadequately connected to information on issues of social relevance.

Review Element 4. Establish a Data Portal to Enhance Access to Water Resource Data Across Agencies

Finding 20: New Data Portals. A new effort by the U.S. Global Change Research Program, using existing agency resources, has been initiated to build an interagency data portal, and this portal also may provide access to hydroclimatic data. The initial phase of this process, conducted with strong coordination among Federal and non-Federal partners, is expected to focus on data generated in support of the National Climate Assessment. The portal also is expected to provide guidance on appropriate use of data and on the use of probabilistic information. Other efforts to better organize data, such as the Integrated Water Resources Sciences and Services activity, also should contribute to enhanced data access.

Review Element 5. Facilitate the Development of Hydrologic and Other Models to Integrate Data That Reflects Groundwater and Surface Water Interactions

Finding 21: Hydroclimatic Statistics and a Changing Climate. Statistical models of hydroclimatic events that are based on the assumption that future conditions will have the same statistical properties as past conditions are not valid. Improved understanding of the statistical characteristics of precipitation and streamflow are needed for effective planning and management under uncertainty.

Finding 22: Integrated Modeling. Although there have been recent advances, current coupled groundwater-surface water models, and fully coupled hydrologic-water quality-ecosystem models are inadequate for forecasting the full range of effects of management actions and/or climate change on the entire resource.

Finding 23: Global Climate Models. There is a need to develop and rigorously test down- scaled Global Climate Models (GCM's) to provide more reliable projections of conditions for for water-resources planning applications, and develop guidance for appropriate applications.

Review Element 6. Apply Hydrologic Information to Resolve Water Resource Problems Including Improvement of Ecological Resiliency

Finding 24: Existing Coordination. Significant ecological and social benefits are expected from enhanced coordination of water-related data collection and integration of activities across the Federal government and with non-Federal partners.

Finding 25: Existing Coordination Mechanisms. Federal agency actions towards improving data integration should continue and be accelerated through the Subcommittee on Water Availability (SWAQ) and the Advisory Committee on Water Information (ACWI). SWAQ should be strengthened to better promote cross-agency communication and additional coordination opportunities. The Advisory Committee on Water Information provides a useful, and perhaps under-utilized forum for collaboration on the collection, storage, management, analysis, and dissemination of climate-relevant water information between Federal and Tribal, state, and local agencies.

Next Steps

Effective management of the Nation's water resources will require meaningful action to address many of the shortcomings that were identified in this report on hydroclimatic observational and modeling systems. Moreover, continued vigilance and effort will be required to sustain and enhance vital observation and modeling systems that currently are in place and operating effectively over a period of years. Federal agencies should consider amending existing programs and policies to incorporate these actions whenever possible and should consider these actions in future budget planning.

This report does not prioritize among the needs that were identified, but prioritization will be required as Congress and the water-resources community begin to deliberate on these findings. Certainly, high priority should be given to maintaining existing capabilities, as measurements of hydroclimatic processes that are not made can never be recovered, creating unrecoverable gaps in the systematic record. Findings in this report that are supported by other documents (for example, the National Monitoring Network and the National Groundwater Monitoring Network) and those groups also deserve priority consideration.

Key next steps for Federal agencies to implement the findings of this report are described below.

1. **Strengthen observational data systems for fresh-water resources and climate change.**
   - Strengthen existing efforts, including the Water Census, through enhancements to the hydroclimatic observational network identified in **Review Element 1** of this report. Information that is critical from a health, safety, and welfare perspective should be given priority, while considering the data needs of land, water and environmental resource managers. Effects on water resources from energy extraction and production, and carbon sequestration (both geologic and biological) need increased emphasis.
   - Conduct ongoing and sustained analyses of hydroclimatic data to identify emerging trends and patterns and to develop new insight into hydroclimatic variability.
2. **Prioritize observational systems that fill important gaps in understanding water supply reliability.**
   - Enhance collection of water-use information, including provision of timely information on withdrawals and return flows (quantity and quality) from surface and groundwater resources and information on withdrawals and consumptive use by sector.
   - Implement the proposed National Streamflow Information Program and the National Groundwater Monitoring Network, both of which are authorized in PL 111-11.
   - Develop and implement a national lake/reservoir level and contents data network.
   - In partnership with private industry, conduct research on new monitoring technologies, including sensors, data transmission, automated quality assurance, and remote-sensing technologies.
3. **Improve water-quality and ecosystem monitoring systems.**
   - Implement the National Water-quality Monitoring Network, which is supported by the Advisory Committee on Water Information and is consistent with the National Ocean Council's Strategic Action Plans.
   - Enhance interagency efforts and support States to monitor and improve mapping of wetland areas and habitat quality on a seasonal basis.

- Implement a waterborne disease tracking network, including all appropriate ancillary data.

4. **Strengthen links between hydroclimatic observational data systems and climate models; improve data management, acquisition, analysis, and reporting.**
   - Link monitoring, observational systems, climate model outputs and other data systems, to update and improve hydroclimatic statistics that support high-priority water management decisions (particularly related to water supply reliability and quality).
   - Build on the initial foundation established by the Integrated Water Resources Science and Services activity to expand and encourage the use of consistent data standards across agencies and with non-governmental partners and other ways to integrate existing data into more comprehensive water information systems.
   - Develop new and improved models (both statistical and deterministic) for assessing hydroclimatic data, developing design conditions, and forecasting likely future conditions for expected scenarios.
   - Develop guidance for water managers on appropriate use of probabilistic projections and model outputs.
5. **Support the establishment of an interagency climate data portal and provide access to high priority water-related datasets.**
   - Promote interagency coordination of diverse data and define the architecture of data systems for freshwater resources to facilitate improved access to these data through a single portal. As a part of the portal, provide for user feedback, including recommendations for improvements or modifications.
6. **Strengthen coordination to improve the quality and accessibility of freshwater data systems including technical outreach and support to stakeholders and decision-makers.**
   - Request that the Subcommittee on Water Availability and Quality (SWAQ) monitor progress on implementing the findings and recommendations of this report and provide annual updates to the National Science and Technology Council and member agencies, and to non-Federal partners.
   - Promote interagency coordination and cooperation to implement the National Water Census (http://water.usgs.gov/wsi/) and Integrated Water Resources Science and Services (National Oceanic and Atmospheric Administration, 2011k).

- Encourage the Advisory Committee on Water Information, an existing Federal Advisory Committee, to establish a new subcommittee or other appropriate mechanism, to solicit and consider input from the public and stakeholders on matters related to freshwater resources and a changing climate and relay these views to Federal water data program managers and the SWAQ.
- Fully engage States, Tribes, local agencies, and interstate organizations, most of whom have water-resources management responsibilities, in the implementation of the findings in this report.

INTRODUCTION

> "We don't yet have the observation networks and capacities that we ought to have to keep track of what's happening on and to the Earth.
>
> . . . That priority on maintaining and expanding the data sets, the observations, the monitoring, is absolutely key. If you don't do that, you can never make up for it, in the sense that we will never know what the Earth was doing in places and times when we weren't monitoring it."
>
> *John Holdren, Science Advisor to the President*
>
> (Eos, 91(51):503-504)

Freshwater, one of the Nation's most valuable natural resources, is under increasing stress from changes in climate, changes in land use and land management, and growing demands for a variety of services. Water supplies and their supporting infrastructure are critical to the health and well-being of society and central to a vibrant economy, food production, energy reliability, and national security. The quantity and quality of freshwater affects the viability of riparian eco-systems, aquatic habitats, and wetlands that support fisheries, wildlife, and recreational activities.

The effects of a changing climate greatly complicates day-to-day management of freshwater and long-term planning for infrastructure to support the Nation's freshwater needs. Documenting, understanding, and predicting the effects of climate change on freshwater supplies and quality remains a significant challenge for the Nation's water-management and scientific community. The consequences of human activities and a changing climate on freshwater already being observed include:

- Changing precipitation patterns (frequency, storm intensity, timing, etc.) that lead to flooding or drought conditions, pollutant runoff from land surfaces into freshwater resources, and disruption of aquatic habitats (for example, Christensen and others, 2007, Intergovernmental Panel on Climate Change, 2007a and 2008);
- Altered snowpack characteristics (depth, density, storage capacity or water equivalent, temperature, melt rate, etc.), that affect timing and volume of snowmelt runoff and the capacity of reservoirs to store water (for example, McCabe and Wolock, 2002);
- Changing sea levels that affect coastal and near-coastal environments, increase impacts of coastal storms on nearshore populations and infrastructure, increase saltwater intrusion into groundwater resources, disrupt water-resources infrastructure, and impair the health of coastal wetlands (for example, Karl and others, 2009); and
- Disruption of healthy aquatic ecosystems due to changing patterns of air and water temperature, precipitation, water quality, land use, and water allocation (for example, Christensen and others, 2007; Intergovernmental Panel on Climate Change, 2007a and 2008; Karl and others, 2009).

A number of agencies and research groups, including the United States Global Change Research Program (USGCRP), are actively engaged in evaluating the effects of climate change on human and natural systems in the United States, including freshwater resources. The recent USGCRP national assessment, released in 2009 (Karl and others, 2009), highlights the importance of providing decision-makers with appropriate data, tools, and models for climate change adaptation. In addition, in its 2010 Progress Report (Council on Environmental Quality, 2010), the Interagency Climate Change Adaptation Task Force (Task Force) called for "improving water resource management in a changing climate." The Task Force, established in 2009 to provide overall direction and guidance to Federal agencies in matters relating to adaptation to climate change, has identified three key actions related to water resource management, one of which is to "strengthen data and information systems for understanding climate change impacts on water."

Congress recognized the importance of supporting water resources decisions when it enacted Section 9506 of the Omnibus Public Lands Act (Public Law 111–11) in March of 2009 (Appendix A). Section 9506 calls for a report to Congress describing the current scientific understanding of impacts of global climate change on the freshwater resources of the United States.

This report was developed in response to the requirements of Section 9506 by an interagency team of water scientists and program managers (Appendix D). The Interagency team cooperated with the Subcommittee on Water Availability and Quality (SWAQ), an interagency subcommittee of the National Science and Technology Council (NSTC), Committee on Environment Natural Resources, and Sustainability (CENRS) (http://www.whitehouse.gov/administration/eop/ostp/nstc/committees/cenrs) and the Interagency Climate Change Adaptation Task Force (http://www.whitehouse.gov/administration/eop/ceq/initiatives/adaptation), and its Water Resources and Climate Change Adaptation Work-group. The interagency team also collaborated with a broad range of interested parties including those represented on the Advisory Committee on Water Information (ACWI) (http://acwi.gov/sogw/pubs/tr/index.html).

Climate-Relevant Water Resource Challenges

Every day, all across the Nation, decisions are made that affect the quantity, quality, and sustainability of water resources. These decisions can have significant economic, health, social, and environmental implications. Some decisions directly affect local water availability such as surface- and groundwater withdrawals and infrastructure development to support drinking water systems, irrigation, energy production, or industrial activities. Decisions regarding land management and development, energy development and fuel extraction, degree of wastewater treatment, and reservoir release schedules also affect water resources. Decision-makers need ready access to reliable data, effective data management, and up-to-date analyses and scientific tools in order to ensure a safe and reliable water supply for humans and ecosystems, today and into the future.

Decisions regarding mitigation of, or adaptation to, effects of climate change also can directly affect water resources. For example, shifts in our national energy portfolio such as increased use of alternative energy sources (for example, biofuels, solar, wind), increased energy efficiencies, or alternative approaches for resource extraction (for example, the use of hydraulic fracturing) can affect local or regional water availability and quality (Electric Power Research Institute, 2002; Mielke and others, 2010). Similarly, the use of carbon capture and sequestration to control greenhouse gas emissions can impact aquifers and surface waters (Dooley and others, 2006; Intergovernmental Panel on Climate Change, 2005). It is, therefore, important to maintain or develop adequate data collection systems that focus on the water related implications of a wide range of decisions and understand the net effect on future water availability and quality (U.S. Environmental Protection Agency, 2010b).

High priority water resources challenges that are affected by climate can be categorized into four interrelated areas:

- Assuring an adequate water supply;
- Protecting human life, health, and property;
- Protecting the quality of freshwater resources and the ecosystems they support; and
- Protecting coastal and ocean resources and the ecosystems they support.

Climate-relevant aspects of each of these challenges are highlighted below. Water managers and decision-makers need access to robust and reliable data, modeling capabilities, and information exchange systems to predict, respond, and adapt to a changing climate.

Assuring an Adequate Water Supply

Federal, State, and local governments, utilities, farmers, energy companies, industry, and private citizens all make decisions that affect and are impacted by the availability, reliability, and quality of water supplies. Critical freshwater needs include:

- Drinking water supplies;
- Agriculture and food security;
- Energy development, production, and generation;
- Industrial cooling and process water;
- Navigation;
- Recreation, and tourism; and
- Healthy aquatic ecosystems.

Water managers and decision-makers need to ensure that water is safely managed and that enough water of the appropriate quantity and quality is available at the necessary time and place to support ecosystems as well as human needs. In addition to climate change, improvements in water-use efficiency, advances in technologies (for example, irrigation, energy, and industry), changes in land-use and development patterns, and changes in the economics of delivering water to end-users need to be considered in water supply decisions.

Table 1. Examples of water resource decisions related to climate

Example of type of decision	Information requirements
Allocating water resources for specific end-uses: • Upgrade existing water systems and develop new supplies to support communities, industry, energy, and agriculture. • Design and operate irrigated and rain-fed agricultural facilities; decide which crops to plant and the extent to which irrigation is needed. • Ensure water availability for reliable power production (hydroelectric, thermoelectric, solar), resource extraction, and other industrial uses (manufacturing, food processing, electronics, biotechnology, etc.). • Ensure sufficient water levels to maintain navigation efficiency and safety. • Ensure adequate water of sufficient quality for endangered species. **Water and related infrastructure:** • Protect the security of our Nation's water infrastructure.	**Water availability:** • Statistical estimates of precipitation frequency, intensity, and duration. • Flood and drought frequency, magnitude, and duration. • Streamflow statistics. • Groundwater storage capacities, recharge rates, and rates of change of groundwater supplies. • Snowpack characteristics. • Glacier characteristics (area, volume, rate-of-change, etc.). • Water-quality conditions. • Sea-level rise patterns and trends. **Water allocation and use patterns:** • Estimates of amount and timing of water withdrawals, demand, and return flows by communities, municipalities, agriculture, industry, and energy production systems. • Uptake and water use requirements for crop production to optimize irrigation scheduling.
• Protect human safety, transportation, and other infrastructure from potential increases in flooding. • Design and operate water storage for agricultural, municipal, and industrial purposes.	• Requirements for quantity and quality of environmental flows and levels to support ecosystem structure and function. • Water rights, interstate agreements, compacts, and court decrees.

Example of type of decision	Information requirements
• Design, operate, upgrade, or rehabilitate degraded water infrastructure (dams, reservoirs, canals, navigational systems, pipelines, storage facilities, drinking water supplies, wastewater and stormwater management, and reuse systems, etc.). • Size and site facilities to manage, process, and convey drinking water, wastewater, return flows, and stormwater. • Design and operate groundwater recharge systems to maximize short- and long-term storage capacity.	• Congressionally authorized reservoir regulation requirements. • Groundwater withdrawal rates. • Cooling water, water use, and wastewater generation rates for thermoelectric power plants and industry (incorporating water footprint of green-house gas mitigation strategies and advances in water-use efficiency), as power demands for cooling increase.

To assure an adequate water supply now and into the future, water managers and decision-makers need the capacity to understand and anticipate:

- The manner in which aquatic ecosystems respond to the short- and long-term changes in freshwater quantity, quality, and availability;
- Historical patterns and projected trends in annual and seasonal variability in streamflows, groundwater levels, and snowpack characteristics to improve water use planning and provisioning;
- The effect of climate, water management decisions, and land-use on water required for agriculture and energy, water and wastewater treatment requirements, human health risks, and aquatic ecosystem health;
- Changes in water quality in response to the types and quantities of chemical and microbial contaminants, warmer and more variable water temperatures, different flow patterns, and alternative water resource management practices (stormwater and rainwater collection and reuse, water-use efficiency, and hybrid systems for centralized and decentralized water reclamation and reuse); and
- Impacts of climate, land-use, and water management practices in conjunction with streamflows, precipitation patterns, and groundwater recharge on saltwater intrusion into freshwater aquifers and estuaries and salinity in water supplies.

Robust and continued measurements are needed to determine the extent to which changes are occurring, the rate of change in both trends and extremes, and the likelihood of crossing thresholds of significance in ecological and hydro-logic systems. The continued, and in some cases, enhanced, collection and documentation of data in a systematic, reliable, and defensible way is critical. As climate, land use, economic, and sociological factors change, data and derivative products form the basis of predictive tools and models to estimate and evaluate factors that affect the amount and quality of water resources likely to be available in the future. Integrated data networks, mapping, and modeling tools provide a basis for predicting potential impacts of climate change at local, regional, and national scales. Equally important are effective protocols and policies for sharing and continually updating the data and tools. Examples of the information needs that underpin water resource decisions are given in table 1.

Protecting Human Life, Health, and Property

Decisions that are made on a daily basis by Federal, State, and local governments, property owners, and insurance industries collectively influence the means by which human life, health, and property are protected from water-related hazards and the means by which the effects of these hazards are mitigated. Key challenges that are relevant to climate change include potential risks associated with:

- Increased potential for loss of life due to changes in the frequency and intensity of precipitation and related flooding;
- Increased incidence of waterborne and vector-borne disease because of changes in water characteristics that affect survival and proliferation of pathogens and vectors;
- Increased cost and energy requirements for producing safe drinking water due to changes in water quality and increased salinity;
- Increased cost and energy requirements for water withdrawals due to lowered groundwater levels and changes in surface water levels;
- Increased potential for drought in some areas leading to altered water quality or quantity;
- Disruptions of power, water, sewer, and emergency services, and failure of flood control and water storage structures, as a result of more extreme rainfall, storm events, and sea-level rise;

- More intensive wildfires in some areas, which affect the quantity and quality of water; and
- Reduced "ecosystem services" within watersheds (ability of the natural system to support outcomes that are economically valued, such as flood control, erosion control, and improvements in water quality that come from filtering by natural systems).

High priority issues for water managers and decision- makers that have been identified in a variety of forums and publications include (for example, Johnson Foundation, 2010):

- Connecting land use and water supply decisions in the context of changing climatic conditions, including implications for ecological impacts;
- Developing early warning systems to prevent and respond to waterborne disease outbreaks;
- Preventing major disruptions to energy supply and water and sewage services (for example, large water main breaks requiring community-wide boil water advisories);
- Identifying areas prone to water inundation to develop appropriate warning, response, and control systems, particularly in urban and developing areas;
- Protecting human safety and economic development through flood mitigation; and
- Assuring adequacy of cooling water for thermoelectric/nuclear power plants.

The types of monitoring and modeling systems needed to protect human life, health, and property are similar to the information needs highlighted in table 1. However, specific attention may be required for documenting incidence of waterborne disease and the proliferation of invasive species in a changing climate. Additional modeling and predictive tools are needed to:

- Predict relations between climate, human activities, and the occurrence and prevalence of waterborne disease vectors and other contaminants in water supplies;
- Forecast changes in water temperature and associated impacts on habitat;

- Understand health risks due to invasive species and microbial pathogens that can survive under changing climate conditions; and
- Identify and safeguard human and ecological populations that are vulnerable to water-related hazards and health-risks.

Protecting the Quality of Freshwater Resources

While considerable progress has been made in reducing pollution within watersheds, the number of impaired water bodies continues to increase (U.S. Environmental Protection Agency, 2011a). Governmental agencies (Federal, State, interstate, Tribal, county, and municipal) and private entities (universities, watershed associations, environmental groups, permitted dischargers, and local citizens) are actively engaged in tracking the occurrence, quantity, quality, distribution, and movement of surface and underground waters (U.S. Environmental Protection Agency, 2009). Significant efforts are being coordinated among the U.S. Environmental Protection Agency (USEPA), the National Oceanic and Atmospheric Administration (NOAA), the U.S. Geological Survey (USGS), the U.S. Fish and Wildlife Service (USFWS), U.S. Forest Service, coastal States, and the National Estuary Programs (NEP) to evaluate ecological and water-quality conditions. For example, The National Water-Quality Assessment Program (NAWQA) (U.S. Geological Survey, 2011a) integrates monitoring data with geographic information on hydrological characteristics, land use, and other landscape features. Examples of other efforts include The National Water Information System (NWIS) (U.S. Geological Survey, 2011b) and the Wadeable Streams Assessment (WSA) (U.S. Environmental Protection Agency, 2006).

Protecting the quantity and quality of freshwater resources and aquatic ecosystems will become increasingly challenging due to the changing patterns of precipitation (intensity, frequency, and type), land use and land-cover change, increasing demands for water, and changing water management practices. Increasing air temperatures will pose direct challenges, including potentially higher evapotranspiration rates, likely leading to increased demand for water for irrigation and energy production. Regulatory agencies, local, State and regional governments, utilities, non-governmental organizations, and industries make decisions that can affect freshwater quality in various ways by:

- Altering local hydrodynamics (through infrastructure projects, water withdrawals, return flows, runoff, or discharges), resulting in changes in runoff patterns, contaminant levels, in-stream flows, and lake and groundwater levels;
- Implementing policy tools [for example, National Pollutant Discharge Elimination System (NPDES) wastewater and stormwater permits, Effluent Limitation Guidelines (ELGs), Water Quality Criteria and Standards (WQS), Designated Uses, and Total Maximum Daily Loads (TMDLs)] for managing impaired waters in accordance with the Clean Water Act;
- Developing and implementing Best Management Practices (BMPs) for managing agricultural and natural resources;
- Restoring and maintaining riparian buffers and flood-plain bottomland hardwood forests to mitigate down-stream flooding and sediment pollution during storm/ flood events;
- Balancing competing water resource needs, including Congressionally authorized reservoir releases, flood risk reduction, and protecting downstream water quality (temperatures and other physical, chemical, or biological water-quality parameters);
- Changing and more variable temperatures, microbial ecology, biogeochemistry, or salinity (due to increased evaporation, discharges of salt-laden constituents, or release of nutrients or other contaminants associated with water use for domestic, industrial, or agricultural purposes);
- Managing water quality changes associated with energy systems (thermoelectric and nuclear power plant cooling), the development of alternative fuel sources (biofuels, shale-gas, tar sands, hydrogen, coal-bed methane), and carbon capture and sequestration;
- Managing and protecting aquifers that are being depleted by groundwater pumping and where aquifers are threatened by coastal subsidence sea level rise, and saltwater intrusion;
- Managing contaminant plumes that are mobilized in groundwater supplies due to increased pumping;
- Preparing for coastal inundation in places where the natural vegetative buffers are ineffective or no longer exist; and
- Addressing sediment loading into reservoirs and developing mitigation and prevention measures to control runoff from intense storms, landscape scale fires, and reservoir operations.

In many cases, water-quality issues are expected to be exacerbated by the impacts of a changing climate on the amount and timing of flows (for example, Larsen, 2010). These changes affect the delivery of contaminants and associated impacts on vulnerable aquatic ecosystems. For example, a large percentage of the total annual loads of nitrogen and phosphorus delivered to coastal ecosystems occurs during a few high-flow events, whereas the toxicity of many compounds is higher during the lowest flows. The extent to which waterborne contaminants (arsenic, fluoride, radionuclides, organics, gases, and microorganisms) are mobilized from aquifer material can vary in response to changes in ground-water levels related to natural and engineered recharge and withdrawal patterns. New health risks may be associated with localized microbial and ecological responses to higher temperature, and in conjunction with changes in contaminant delivery, will impact the population dynamics and survival of invasive species, pathogens, and algae.

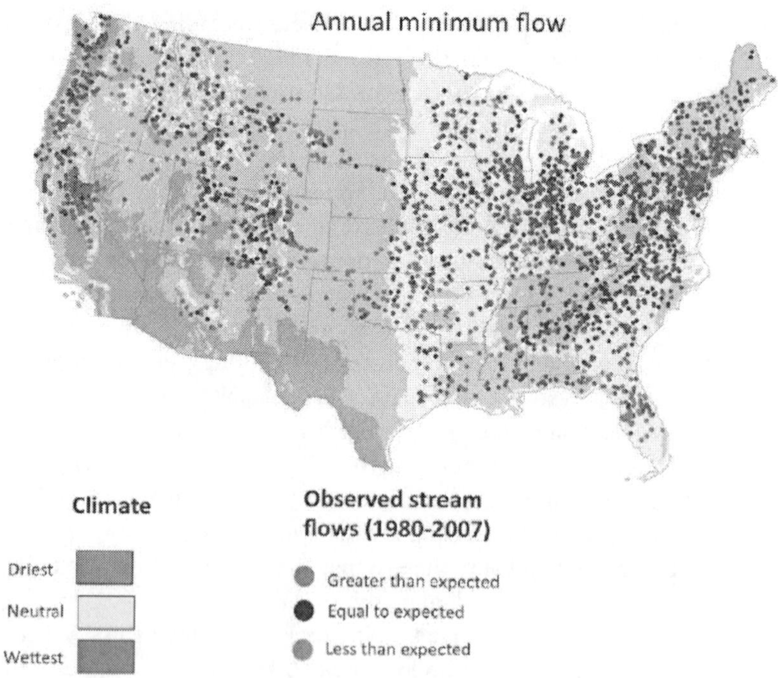

Figure 1. Alteration of minimum streamflow from expected minimum streamflow at 2,888 sites (Carlisle and others, 2010).

Changes in the natural flow regime are an important contributor to degraded river ecosystems and loss of native species (Carlisle and others, 2010; Poff and Zimmerman, 2010). In a national study at 1,059 reference (unimpacted) and 2,888 other sites, (Figure 1) Carlisle and others (2010) compared maximum and minimum flows at the reference sites to those that would be expected to occur at the other sites. The extent of streamflow alteration from expected conditions was the primary predictor of biological integrity at all sites.

Protecting Coastal and Ocean Resources

Some of the most densely populated areas in the Nation are in the coastal regions. These regions play an integral role in the economy, supporting community well-being, commerce (ports), transportation, energy (oil, natural gas, refineries, wind, tidal and wave energy, solar), fisheries, tourism, and recreation. Coastal waters are valued for their ecological richness and encompass many unique habitats (estuaries, coastal wetlands, seagrass meadows, coral reefs, mangrove and kelp forests, and upwelling areas). Coastal ecosystems yield highly productive fisheries and provide breeding habitat for 85 percent of U.S. waterfowl and other migratory birds (U.S. Environmental Protection Agency, 2011b).

Coastal and ocean resources and coastal communities face unique challenges associated with a changing and changed climate (Nelson and others, 2011). Maintenance of healthy coastal resources is directly dependent on the fresh-water systems that sustain coastal environments. For example, many species are sensitive to temperature and salinity, which are directly influenced by weather patterns and freshwater inputs. Temperature and salinity also can influence the frequency and persistence of algal blooms (red tides, cyanobacterial blooms, pfiesteria, etc.).

Key coastal issues that are likely to be impacted by climate change include:

- **Extreme events:** Changes in the intensity, duration, and number of hurricanes, tropical storms, and intense coastal low pressure systems may increase coastal hazards and affect portions of the coast previously unimpacted by these systems.

- **Emergency Response Infrastructure:** Changes in hazards as hurricanes and other coastal storms increase in intensity and shoreline erosion continues (Hapke and others, 2010);
- **Drinking water quality:** Changes in the condition of coastal freshwater aquifers and drinking water supplies that are vulnerable to risks as sea levels rise or salt water intrusion increases;
- **Sea level changes:** Rising sea levels and changes in storm frequency, intensity, and duration that affect the design, operation, and management of water infrastructure (drinking water systems, water storage, waste-water and stormwater management and reuse) (for example, Intergovernmental Panel on Climate Change, 2007b);
- **Coastal habitats:** Alterations in estuarine and coastal habitats at risk of inundation as sea levels rise (Inter-governmental Panel on Climate Change, 2001);
- **Hypoxia:** The extent of hypoxia (lowered oxygen levels) and "dead zones" in the Gulf of Mexico, Chesapeake Bay, and other water bodies as contaminant loads change; and
- **Acidification:** The quality of coastal and ocean fisheries and habitat, such as coral reefs, at risk due to warmer waters and ocean acidification.

The National Assessment of Coastal Vulnerability to Sea Level Rise has classified United States coastal environments to develop a comparative index of the potential for physical shoreline changes in response to sea level rise (Figure 2; Thieler and others, 1999–2000). High-vulnerability areas tend to be barrier islands with small tidal ranges, large waves, a low coastal slope, and high historical rates of sea-level rise.

The recent National Coastal Commission Report (NCCR) (coordinated through the USEPA, NOAA, USGS, USFWS, coastal States, and the NEPs) rated the overall condition of coastal environments as fair (U.S. Environmental Protection Agency, 2008). From a water-quality standpoint, the National Coastal Commission develops a net score based on five indices of ecological condition (U.S. Environmental Protection Agency, 2008):

- Water-quality index (including dissolved oxygen, chlorophyll a, nitrogen, phosphorus, and water clarity);
- Sediment quality index [including sediment toxicity, sediment contaminants, and sediment total organic carbon (TOC)];
- Benthic index;

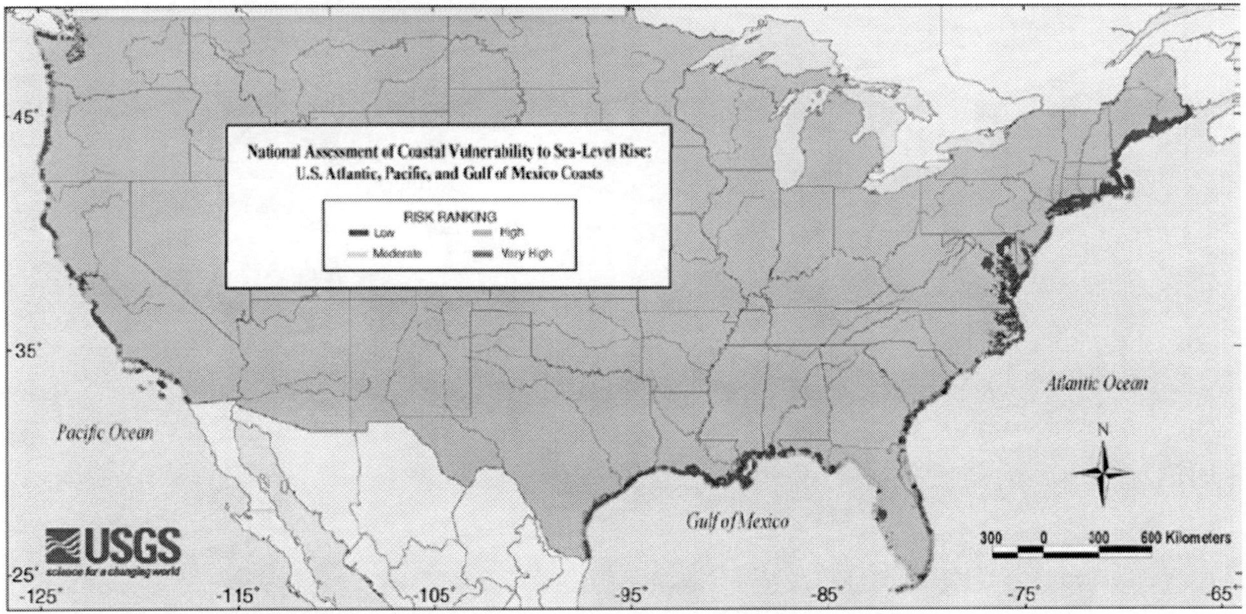

Figure 2. Relative vulnerability of coastal areas to seal level rise (Thieler and others, 1999–2000).

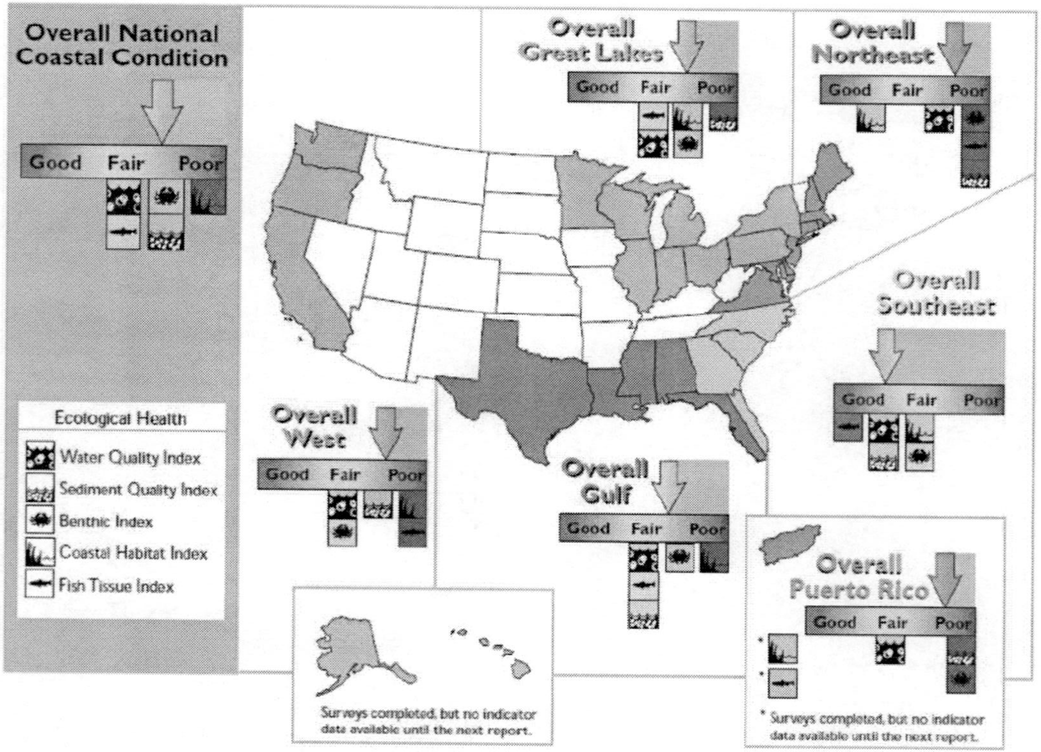

Figure 3. Overview of coastal condition scorecard (U.S. Environmental Protection Agency, 2008).

- Coastal habitat index; and
- A fish tissue contaminants index.

An overview of the national coastal condition score-card is shown in figure 3. As of 2008, 18 percent of estuarine waters were considered to be impaired for fishing, based on the prevalence of fish tissue contaminants (U.S. Environmental Protection Agency, 2008).

Examples of up-to-date climate-relevant information, data, maps, and models that are needed to make decisions about the management of coastal and ocean resources and the ecosystems they support include:

- Estimates of relative sea level change and storm-surge hazards to determine how tidal fluctuations and sea-level rise will impact coastal environments, communities, and their water infrastructure;
- Geospatial data on the locations and characteristics of wetlands and their resiliency to sea-level rise to predict the effects on coastal wetlands and adjacent coastal resources of a changing climate;
- An understanding of coastal ecosystem responses to changing upland hydrology, flood frequencies, and water-quality conditions; and

Hydrodynamic models to predict the likely extent and consequences of saltwater intrusion under various conditions and water management practices (groundwater recharge, water allocations, ocean outfalls, and deep well injection).

REVIEW ELEMENTS

Section 9506 identifies six interrelated review elements (Appendix A):

- Assess adequacy of current hydro-climatological observation networks;
- Identify data gaps in current water monitoring networks;
- Improve data management to increase utility of information that is collected and efficiency of data acquisition and reporting;
- Establish a data portal to enhance access to water resource data across agencies;

- Facilitate the development of hydrologic and other models to integrate data that reflects groundwater and surface water interactions; and
- Apply hydrologic models to resolve water resource problems including improvement of ecological resiliency.

Each review is discussed in turn, using information contributed from agencies and groups that produce data, manage networks, develop and maintain models, and operate data portals, or have plans to do so. It is important to recognize that each of the six review elements encompass significant ongoing activity both inside and outside of the Federal government. Because of the general overlap of elements 1 (adequacy of networks) and 2 (gaps in current networks), these two elements are combined for the purposes of this review.

There is substantial complexity in monitoring and observing systems and the manner in which data from these systems are combined to increase understanding of past, current, and future conditions and trends. Some of the interconnections among the review elements are illustrated in figure 4. As shown, networks are used to organize and map individual data elements. Integrated networks can be compiled and shared through data portals. Models require data for development and application. Mapping and modeling tools build upon historical data to evaluate past trends and to project the range of outcomes of future climate, land-use, and water-management scenarios on water resources and the human and aquatic systems they support. The extent to which the physical hydrologic system, as well as observations, data, modeling, mapping, and decision-making are interconnected highlights the importance of coordination, collaboration, and communication. Some initial efforts to integrate data are underway through a variety of formal and *ad hoc* groups, such as the USGCRP, the National Water-quality Monitoring Council, and the Consortium for the Advancement of Hydrologic Sciences, Inc. (CUAHSI).

Background

Numerous reports from a wide range of organizations and constituencies have called for strengthening the Nation's environmental monitoring networks. The Congressional Research Service (2009), in a review of 1973 National Water Commission findings, noted that "Few would disagree with or reject a recommendation for more and better water-quality monitoring . . . environmental monitoring generally, and water-quality monitoring specifically, receive less priority and funding than do regulatory or capital improvement programs." Tracy Meehan, The Cadmus Group, has said, "Data and water monitoring, both as to quantity and quality, are foundational to water management especially in an era in which the focus has moved away from an almost exclusive preoccupation with end-of-the-pipe point source dischargers to a broader concern with watersheds, generally, and the physical and biological integrity, not just chemical, of our nation's aquatic resources (Meehan, 2010)."

> "The only way to figure out what is happening to our planet is to measure it, and this means tracking changes decade after decade and poring over the records. A point of diminishing scientific returns has never been realized in what is now known as the "Keeling Curve," the Mauna Loa CO_2 record."
>
> *Ralph Keeling, 2008*

The Johnson Foundation (2010), recognizing the role of science in water-resources management called for the generation of "sound science that accounts for the dynamic nature of freshwater systems and our emerging understanding of climate change impacts on water that can be shared in real-time to inform mitigation and adaptive management strategies." Furthermore, the Foundation recommended "that Federal agencies, especially those within the U.S. Department of the Interior, expand existing nationwide freshwater quality and quantity monitoring and data collection networks and outfit them with cutting-edge technology that enables rapid data analysis and real-time data sharing. The installation of additional streamgages, water meters, groundwater monitoring wells, and better estimates of consumptive use are of paramount importance for the effective management of available water supplies." The Water Environment Federation (2010) notes that "Water resource managers must also adapt to new information, consistently refining their approach to sustainability. They must continually incorporate locally relevant data into long-range planning and recognize that climate change is altering long-held beliefs about hydrological norms."

The Climate Change and Water Working Group (CCAWWG) (National Oceanic and Atmospheric Administration, 2008) is an interagency group that has been working to identify critical gaps relevant to climate change adaptation. Several recent reports (Brekke and others, 2009, 2011) have highlighted the need for collaborative efforts across the water management and scientific communities to develop, test, and apply new methods, tools, and capabilities. Key goals include (1) enhanced monitoring and data collection, and (2) improved decision-making under uncertainty.

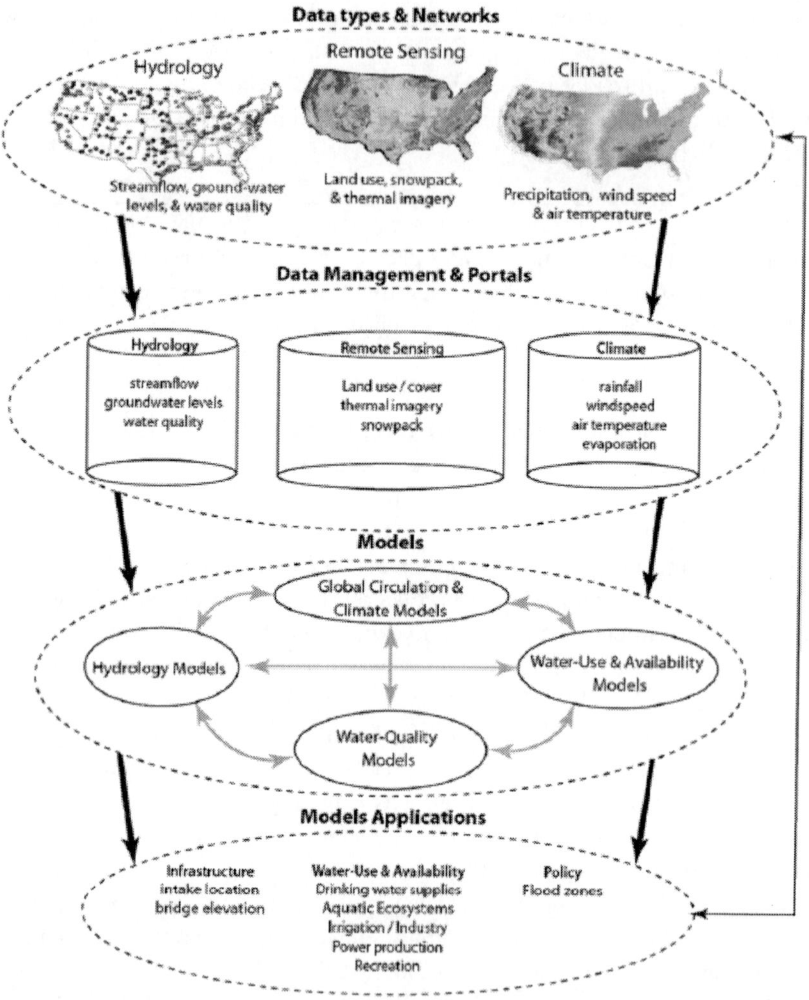

Figure 4. Examples of interconnections among data, models, and model applications for decision-makers.

In an October 2010 report to the President, the Interagency Climate Change Adaptation Task Force (2010) recommended that water resources management be enhanced in response to a changing climate by strengthening data and information systems. This report also called for development of a *National Action Plan* addressing management of freshwater resources in a changing climate. The draft *National Action Plan* (http://www.whitehouse.gov/administration/eop/ceq/ initiatives/adaptation) describes a number of actions related to improvement of data and information for adaptation to climate change.

There also is wide recognition that better coordination of environmental data collection would benefit the Nation. The U.S. Governmental Accountability Office (GAO) (2004) noted that 15 agencies collect water-quality data, and that hundreds of other organizations collect these data, in contrast to the relatively few agencies that collect water quantity data. GAO identified several barriers to improved coordination, including data-quality objectives, methods, awareness of others collecting the same type of data, and low priority of coordination. The Subcommittee on Groundwater (2009) addressed the issue of coordination by proposing a mechanism whereby groundwater data collected by various entities could be readily exchanged through the use of common definitions and data catalogs.

In May 2011, the National Oceanic and Atmospheric Administration, U.S. Army Corps of Engineers, and U.S. Geological Survey signed a memorandum of understanding better align agency programs to support integrative and adaptive water management. The partnership will address the goals of the Integrated Science and Services initiative (http:// nws.noaa.gov/oh/docs/IWRSS_1p_summary.pdf), with the objective of building a Federal Support Toolbox for integrated water-resources management. The U.S. Geological Survey Water Census (http://pubs.usgs.gov/fs/2007/3112/fs2007-3112. pdf) will support this initiative, and will work across Federal, State, Tribal, and local agencies to integrate water-resources information with the goal of developing detailed water budgets for most river basins in the U.S.

Adaptive management has emerged as one approach to managing water resources and the ecosystems they support in the face of a variety of uncertainties, including climate change. Endangered species are being managed adaptively in Upper Colorado River, the Delaware River, the Missouri River, the San Joaquin/Sacramento Bay-Delta, the Platte River, and elsewhere. Adaptive management can only be successful when supported by a robust data-collection and analysis program, or as the Council on Environmental Quality (2003) suggests, "monitor and adapt."

Review Elements 1 and 2: Assess Current Hydroclimatological Observation Networks and Identify Data Gaps

Review Element from Section 9506:

> "to assess the extent to which the conduct of measures of streamflow, groundwater levels, soil moisture, evapotranspiration rates, evaporation rates, snowpack levels, precipitation amounts, and glacier mass is necessary to improve the understanding of the Federal Government and the States with respect to each impact of global climate change on water resources…"

and

> "to identify data gaps in current water monitoring networks that must be addressed to improve the capability of the Federal Government and the States to measure, analyze, and predict changes to the quality and quantity of water resources, including flood risks, that are directly or indirectly affected by global climate change…"

For simplicity of presentation, review elements 1 and 2 have been subdivided into six sub-elements:

- Streamflow;
- Groundwater levels;
- Soil moisture and evapotranspiration;
- Precipitation amounts, snowpack, and glacier mass;
- Water quality and ecosystem conditions; and
- Water withdrawals and consumptive use.

Finding 1: Broad Support for Improved Monitoring Systems. There is broad and continuing support from a wide constituency to sustain and strengthen the Nation's hydroclimatic monitoring systems, including research to support those systems. In particular, adaptive management and water-resources planning tools are only effective when supported by strong data networks.

Significant efforts are devoted to the measuring and monitoring of various attributes of water resources. In some cases, however, there is a lack of coordination among the various efforts because of differing data-collection purposes, methods, and data-quality objectives. Differences in the content and interoperability of data networks reflect current and historic funding levels, programmatic history, agency culture, technology, and the priorities of the organization(s) that collect and manage the data. Regardless of these differences, scientists and decision-makers often combine disparate data sets to develop models and conduct analyses necessary for the management of the Nation's water resources.

Streamflow

Surface waters (streams, rivers, lakes, and reservoirs) serve as a source of drinking water to more than 170 million Americans and supply over 80 percent of the water withdrawals in the United States (Kenny and others, 2009). Much of the Nation's treated municipal, domestic, and industrial water is returned to surface waters along with irrigation return flows, and stormwater runoff. Surface waters provide important aquatic habitats and maintain the viability of commercially valuable fisheries, protected species and aquatic ecosystems. Surface waters also generate tremendous economic benefit through support of agricultural irrigation, tourism (fishing, canoeing, rafting, etc.), navigation, power production, and other industrial freshwater needs. However, changes in the natural flow and thermal regime of rivers, coupled with water-borne contamination have degraded the availability and quality of critical water resources and the ecosystems they support (U.S. Environmental Protection Agency, 2010b).

Until recently, it was assumed that future streamflow conditions (flow magnitude, seasonal distribution of flows, occurrence of floods and droughts, etc.) could be estimated from historical data. The idea that the past streamflow conditions are a proxy for the future is based on the statistical property of stationarity (Milly and others, 2008). However, as natural systems respond to changing climate and other types of global change (for example, land and water management practices), and as we gain improved understanding of the inherent long-term variability in natural systems, there is increasing uncertainty about estimates of future streamflow conditions obtained using currently available models and statistical methods. Nevertheless, continuous, long-term records of stream-flow are indispensable to the estimation of future streamflow conditions, but the number of long-term streamflow gages continues to decline (Figure 5).

The USGS operates more than 7,500 streamgages (Figure 6; Norris, 2009) in cooperation with about 850 Federal, State, Tribal, and local partners. In most cases, streamflow observations are made at 1-minute to 1-hour intervals

(U.S. Geological Survey, 2011b) and are transmitted via satellite to Federal [for example, National Weather Service (NWS)] and local users using the Internet. The immediate availability of stream-flow data (often reported within 1 hour of collection) helps decision-makers respond to emergency situations and also address day-to-day management of irrigation systems, water supply and wastewater facilities, reservoirs, canals, and navigation systems. For example, NWS River Forecast Centers (RFC's) (U.S. Geological Survey, 2011c) use streamflow data to provide short-term (about 7–day) flood forecasts, as well as monthly to seasonal water availability forecasts. The compilation of long-term streamflow datasets (U.S. Geological Survey, 2011c) provides aggregated data that are used to design water systems and infrastructure, calibrate forecast models, and evaluate streamflow response to climate change (McCabe and Wolock, 2002).

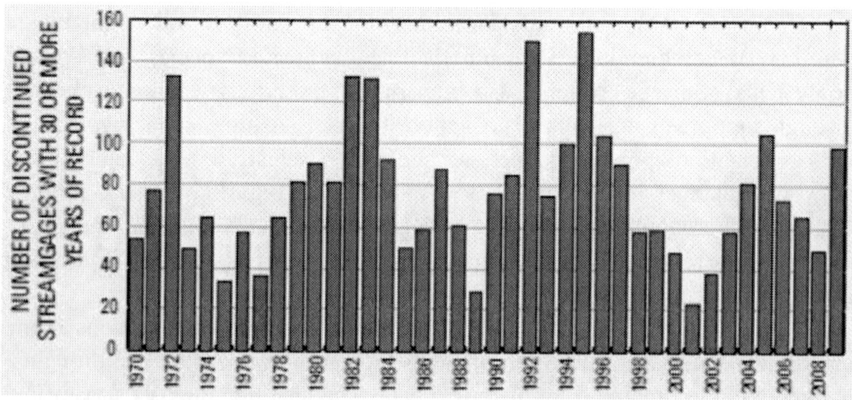

Figure 5. Number of discontinued U.S. Geological Survey streamflow gages with 30 or more years of record, 1970–2009 (Norris, 2009).

Almost half of the funding for the current streamgage network comes from State, regional, local, and Tribal partners (Norris, 2009). As a result, the availability of funding tends to fluctuate from year to year, leading to instability in the network.

The National Streamflow Information Program (NSIP) (U.S. Geological Survey, 2011d) was initiated in 2003 in response to Congressional and stakeholder concerns about (1) the decrease in the number of operating streamgages, including a disproportionate loss of streamgages with a long period of record (Figure 5); (2) the inability of the USGS to continue operating high-priority streamgages in an environment of reduced funding through partnerships; and (3) the increasing demand for streamflow information due to emerging resource-management issues and new data-delivery capabilities (Norris, 2009).

In 2009, NSIP was authorized as part of Section 9507 of PL 111-11, with a full implementation goal of 2019. The National Streamflow Information Program was begun in 2003 with 5 program goals (U.S. Geological Survey, 2011e). These goals are to (1) develop a stable streamgaging network; (2) conduct intensive data collection during floods and droughts; (3) conduct regional and national streamflow assessments; (4) improve data delivery; and (5) invest in methods development research. Current progress toward meeting the goals include 15 percent of the stable streamgaging network in place with some improvements in data delivery.

In addition to the USGS network, a number of smaller streamgage networks are operated by other Federal agencies and States. The U.S. Army Corps of Engineers (USACE) operates a network of gages, many of which provide water level only, in support of their civil works programs. Information generally is available by Corps District (for example, St. Louis district (http://mvs-wc.mvs.usace.army.mil/dresriv.html; http://www.mvp-wc.usace.army.mil/dcp/), or for all Corps sites (http://rivergages.mvd.usace.army.mil/WaterControl/new/layout.cfm).

Similarly, the U.S. Fish and Wildlife Service collects streamflow information on some of their refuges (for example, http://alaska.fws.gov/water/monitor.htm), as do the National Park Service (NPS) (http://www.nature.nps.gov/ water/), and the U.S. Forest Service (for example, see Water at http://www.fs.usda.gov/rds/archive/datacatalog/), each for agency-specific purposes. Some States also operate their own streamflow networks. For example, Colorado maintains a network of gages to assist in administration of water rights (http://www.dwr.state.co.us/SurfaceWater/Default.aspx).

The West Virginia Water Gaging Council maintains a web page from which users can access streamflow, flood warning, precipitation, and groundwater data collected by a variety of agencies and organizations (http://wvwgc.wvca.us/).

Tremendous benefits could be gained for the Nation if these disparate data sets were more broadly available and included information on methods of collection, quality, and accuracy.

Finding 2: Streamflow Data. The National Streamflow Information Program was begun in 2003 with 5 program goals. These goals are to (1) develop a stable streamgaging network; (2) conduct intensive data collection during floods and droughts; (3) conduct regional and national streamflow assessments; (4) improve data delivery; and (5) invest in methods development research. Current progress toward meeting the goals include 15 percent of the stable streamgaging network in place with some improvements in data delivery.

The amount of water stored in lakes and reservoirs is directly linked to streamflow because of the effect of dam releases on flows, as well as losses from incoming streamflow to evaporation. Reservoir lake-level data, when converted to water volume, provide important information to water managers regarding availability of short-term (days to months) future water supplies, dam operation, downstream navigation, and maintenance of instream flows.

Finding 3: Streamflow Information Coordination. Streamflow information is collected by a variety of Federal and State agencies, generally for agency-specific purposes. Tremendous benefits could be gained for the Nation if these disparate data sets were more broadly available, interoperable, and included information on methods of collection, quality, and accuracy.

The USACE (http://www.ndc.iwr.usace.army.mil/), and the Bureau of Reclamation (BoR) (http://www.usbr.gov/lc/ region/g4000/hourly/levels.html) provide lake level data on most of the reservoirs which they operate, with some of the lake level measurements made by the USGS. Similar information is reported by the Tennessee Valley Authority (TVA) (http://www.tva.com/ river/lakeinfo/index.htm). Data are available at varying intervals, from hourly to monthly.

Although Federal projects (USACE, TVA, BoR) account for the majority of water storage in the U.S., a number of river systems are managed by private entities (for example, the Catawba–Wateree River in North and South Carolina, managed by Duke Energy, and the Alabama River, managed by Alabama Power). Numerous small impoundments are managed by the Natural Resources Conservation Service (NRCS), irrigation districts, and water utilities, some of which have continuous lake level information. However, limited data exist for many natural lakes.

Water resource managers also need knowledge about reservoir and lake volumes in conjunction with water-level monitoring in order to estimate reservoir contents at any particular time. Sediments tend to accumulate in lakes and reservoirs, reducing the volume available for storing water. In addition to affecting the available volume, trapped sediments in reservoirs can be an important sink for carbon and also may harbor contaminants and pathogens (Stallard, 1998). Precipitation patterns (form, intensity, and frequency), fires, intense storms and droughts, along with changes in hydrology, land use, and land management affect the amount of sediment transported into reservoirs. The Reservoir Sedimentation Database (RESSED) (U.S. Geological Survey, 2009b), started in 2009, is a preliminary effort to develop a dynamic database of reservoir volume. Bathymetric reservoir data are used to estimate the volume of the underwater environment. NOAA provides bathymetric data on coastal waters (National Oceanic and Atmospheric Administration, 2011e) and collaborates with the Canadian Hydrographic Service (CHS) to provide data on the Great Lakes, but measurements of other freshwater systems have declined substantially (figure 7; Ackerman and others, 2009).

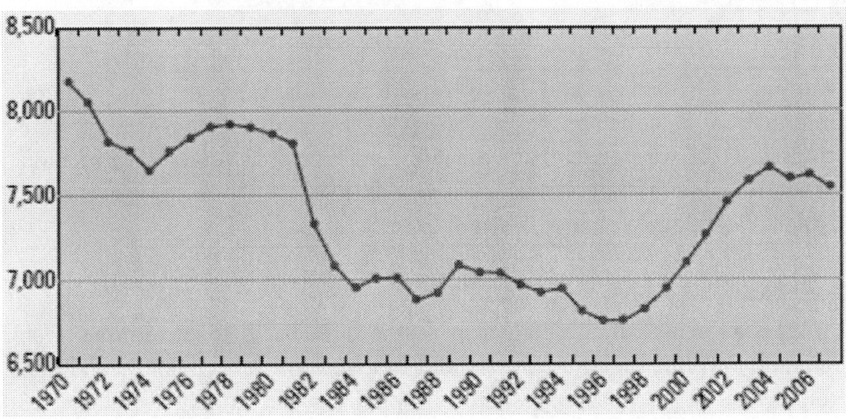

Figure 6. Number of active streamgages in U.S. Geological Survey network, 1970–2007 (Norris, 2009).

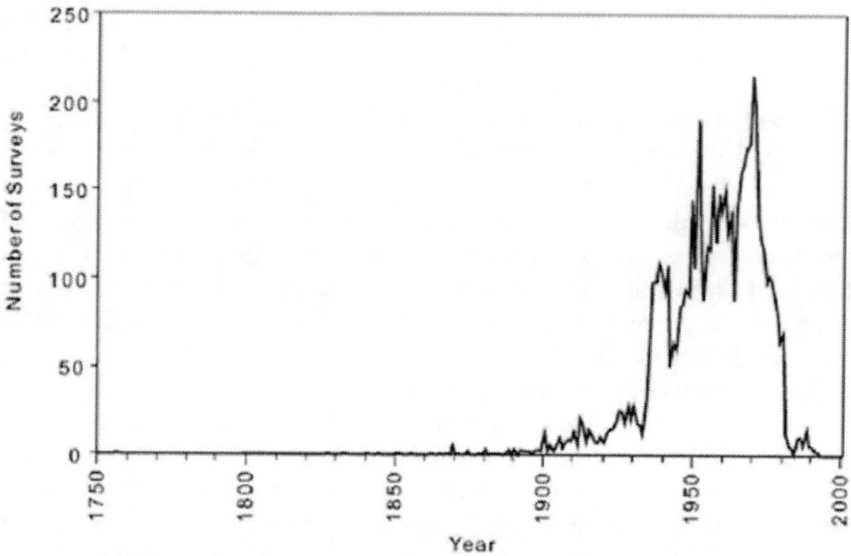

Figure 7. United States reservoir sedimentation surveys, 1750–2000 (Ackerman and others, 2009).

> **Finding 4: Reservoir and Lake Data.** Reservoir water-level and volume information is collected by a variety of Federal agencies and private entities. Data are not widely available, and reservoir volume information generally is out of date. No data are available for many natural lakes, and a national lake level network does not exist.

Groundwater Levels

Measurements of groundwater levels are used to determine water availability within specific aquifers and to assess overall climatic conditions. Groundwater is the primary source of public water supply for about 130 million Americans, including about 97 percent of the rural population (U.S. Geological Survey, 1995; Dennehy, 2005). Approximately 42 percent of irrigation water supplies are withdrawn from groundwater (Hutson and others, 2004). Additionally, ground-water sustains and replenishes rivers and streams during the hottest and driest parts of the year, providing critical flow to aquatic ecosystems. Groundwater accounts for a large percentage of the total flow in most streams, with as much as 80 percent of streamflow originating from groundwater in some parts of the Nation (Winter and others, 1998). As locations for additional reservoirs become limited and availability of water from the Nation's streams becomes increasingly scarce, withdrawals from groundwater are likely to increase (Hutson and others, 2004); even brackish groundwater is seen as a potential future water supply. In many areas, including California and the southeastern U.S., increases in groundwater use during surface water shortages have been dramatic. Data and models are needed to better understand the implications of precipitation variability, and withdrawals on groundwater availability and, because of the tight coupling of surface and groundwater, on streamflows and aquatic ecosystems.

Climate change, changes in agricultural practices, and energy development likely will introduce other impacts on groundwater systems, with cascading effects on the rate and intensity of pumping to support local water needs (irrigation, frost prevention, fire-fighting, urban uses). For example, intensive short-term demands on groundwater within agricultural regions in response to freezing or drought conditions has led to rapid lowering of instream flows and aquifer levels with sub-sequent generation of sinkholes, particularly in karst regions (Tihansky, 1999). Climate change adaptation and mitigation efforts also can spur development of competing water uses in a region for activities such as biofuel production, hydraulic fracturing, mineral and fuel resource extraction, industrial activities, or agriculture. Recent studies have suggested that groundwater depletion is contributing to sea level rise through aquifer compaction and land subsidence (Wada and others, 2010).

The USGS maintains and monitors more than 20,000 observation wells in cooperation with State and local agencies, with monitoring frequencies ranging from hourly to annually, or less infrequently. Industries and municipalities also operate and maintain wells. In addition to data on water levels, water quality and precipitation data are collected at many sites. Observation and monitoring wells also are used to evaluate underground sources of drinking water, to support the Underground Injection Control (UIC) program under the Safe Drinking Water Act (SDWA), for mineral and energy resource extraction, to develop aquifer storage and recovery or ground-water recharge programs, and for groundwater remediation projects. Nevertheless, according to the U.S. Governmental Accountability Office (2004), no Federal agencies are collecting groundwater level data on a national scale.

A complete inventory of all active (or abandoned) wells does not exist. In fact, there is wide variability in the amount of information available, depending on how active State governments and local water districts are in managing ground-water rights and well drilling activities. Although a complete inventory is not a necessity for effective water-resources management, such an inventory would be useful for protecting groundwater resources and for designing data collection networks. Certainly, information on all future wells should be made available through a common framework. Limited analyses of the interrelationships among water quality, water levels, and well pumping practices have been conducted. These types of analyses generally require application of complex ground-water models founded on good data and a reasonable representation of the hydrogeology.

Over the past few years, concerted efforts have been made to develop a Climate Response Network (CRN, which is distinct from NOAA's Climate Reference Network) to measure response of the groundwater system to climate variations (U.S. Geological Survey, 2011f). The CRN is designed to monitor groundwater fluctuations in unstressed aquifers, as these systems often can provide early indications of drought. While modest progress has been made, the CRN currently includes sites in only 60 percent of the Nation's 42 principal aquifers (U.S. Geological Survey, 2003) and in fewer than half of the 366 NOAA climate divisions in the United States and Puerto Rico (Cunningham and others, 2007). In addition, when multiple aquifers are present in a given climate division there is a need to provide additional monitoring sites. Such a network, one climate response well per principal aquifer in each climate division of the U.S., might be considered a minimum design for a CRN. The CRN, however, is not designed to monitor the effects of groundwater withdrawals on groundwater availability.

Section 9507 of PL 111-11 directs the Secretary of the Interior to "develop a systematic groundwater monitoring program for each major aquifer system located in the United States." The legislation contains a number of key features, including "expanding the network of monitoring wells to include each climate division." Appropriations for the ground-water monitoring network are authorized for 2009–2023. The Subcommittee on Groundwater (SOGW) of the Federal Advisory Committee on Water Information recently completed a design for a National Ground-Water Monitoring Network (Subcommittee on Groundwater, 2009). The proposed network is a compilation of wells from existing State and Federal monitoring activities, and was designed to focus on monitoring the most productive aquifers. The final design will vary by aquifer, but likely will specify a minimum number of wells to achieve a desired monitoring density. The network would monitor both stressed and unstressed systems (Figure 8), with targeted monitoring in areas of concern (areas of contamination or depletion). The National Network design also calls for development of a data-management system with applications that facilitate the retrieval and display of data in dispersed databases. The network currently (2011) is being piloted in six States (Advisory Committee on Water Information, 2011): Illinois-Indiana, Minnesota, Montana, New Jersey, and Texas.

Finding 5: Groundwater-Level Monitoring. The Advisory Committee on Water Information's Subcommittee on Groundwater has developed a comprehensive national groundwater monitoring network that could provide data and information necessary for planning, management, and development of the Nation's groundwater resources in a sustainable manner. This network was authorized in Section 9507 of PL 111-11.

Soil Moisture and Evapotranspiration

Soil moisture is critical for agricultural crop production. With climate change and the concomitant changes in the length of the growing season, farmers' water use practices will need to adapt to changes in crop requirements, available precipitation, and the rate of evapotranspiration (Intergovernmental Panel on Climate Change, 2007c). The amount of moisture stored in the soil is directly related to the balance between water saturation (through rainfall or irrigation) and evapotranspiration, or the amount of water released to the environment through the growth of plants, trees, and vegetation. These factors, in conjunction with land use patterns and local hydrology influence the rate and extent of groundwater recharge and generation of runoff.

Changes in soil moisture are likely to vary, depending on watershed characteristics, soil properties, temperature, and the size and the role of the landscape in storing water (Furniss and others, 2010). The local effects of climate change on soil moisture are complex and interdependent on multiple factors. Soil characteristics dictate the relative capability of a soil to store water, influencing the magnitude and persistence of soil moisture. Climate change can affect soil moisture through changes in precipitation and air temperature. The quality of irrigation water, particularly the salt and mineral content, also affects short-term and long-term soil moisture. Infiltration capacity and water-holding capacity of many soils are influenced by the frequency and intensity of freezing. The seasonality of soil moisture is important for sustainable agriculture and often is the controlling factor in whether irrigation water needs to be applied to produce crops. For example, in the West and Mid-west, soils are likely to dry earlier in the year with increasing air temperatures, imposing stress on existing vegetation and leading to changes in the diversity of native vegetation and insect outbreaks. Changes in the timing of snowpack runoff also may affect the seasonality of water availability. Changes in the length and timing of the frost-season affect the survival and proliferation of insect pests and the migration patterns of wildlife, all of which affect ecosystems and soil vitality. Increased decomposition rates from higher temperatures will deplete soil organic matter more rapidly (Furniss and others, 2010), and changing air and soil temperatures will affect the rate of consumptive water use.

Soil moisture is a significant input parameter for meteorological, drought and rainfall-runoff forecasts. The 2,200 interagency RAWS (Remote Automated Weather Stations) network includes measurements of soil moisture at some stations (Remote Automated Weather Stations, 2011). The NRCS Soil Climate Analysis Network (SCAN) contains 180 sites in 40 States and U.S. Territories at which soil moisture is measured at 5 depths below the land surface (U.S. Department of Agriculture, 2011a). About nine other meteorological parameters also are measured at the sites, and the data collected at SCAN sites provide all of the information required to calculate a reference evapotranspiration value at each site. SCAN coverage is non-uniform throughout the country with some States having only one site, and other States having 10 or more sites. In addition to organized networks, there likely are numerous private soil-moisture data-collection sites operated by landowners to schedule irrigation. These networks are not connected in any useful manner.

The U.S. Climate Reference Network currently consists of 114 stations that record temperature, precipitation, solar radiation, surface skin temperature, and surface winds. These stations are being expanded to track soil moisture and soil temperature at five depths (5, 10, 20, 50, and 100 cm) (National Oceanic and Atmospheric Administration, 2011a). The goal is to implement about 30 additional stations over the next 5 years and provide calibration sites for the National Aeronautic and Space Administration's (NASA) Soil Moisture Active Passive (SMAP) mission, planned for 2014 (National Aeronautic and Space Administration, 2011b). Nevertheless, a comprehensive design for a national soil moisture network has not been completed. The SMAP mission is designed to provide soil moisture data at daily intervals, with a horizontal resolution of about 10 km x 10 km. The SMAP mission, in conjunction with a robust *in situ* network for calibration, could provide adequate soil moisture data for many applications at regular intervals over much of the country.

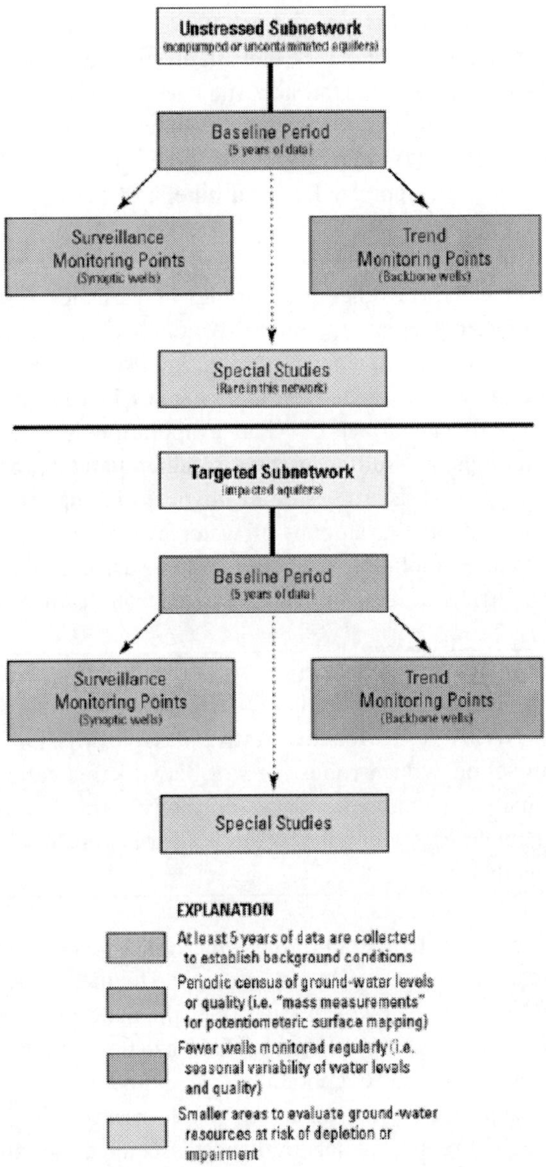

Figure 8. National Groundwater Monitoring Network design, showing types of networks and relations among networks (Subcommittee on Groundwater, 2009).

Evapotranspiration (ET) is the second largest component of the water cycle after precipitation, and ET is one of the primary consumers of solar energy at the land surface. Thus, an important feature of ET is that the process couples the energy budget and the water budget at the land surface. This coupling makes it difficult to correctly model ET in downscaled global climate simulation models (for example, Bae and others, 2011; Mueller, and others, 2011).

ET is very sensitive to a change in land cover and weather. Change in essentially any one of the other meteorological variables (precipitation, air temperature, humidity, solar radiation, wind, barometric pressure, etc.), changes in the character (frequency, duration, intensity, type) of precipitation, changes in vegetation and land cover, and changes in vegetation response to greenhouse gases all can affect ET. Pan evaporation is measured at many NWS weather stations, as well as at some weather stations operated by other agencies. Pan evaporation is a measure of the potential maximum evaporation if there was no limit on the amount of water available. Pan evaporation is converted to actual evaporation from free water surfaces (that is, lakes and ponds) using a coefficient, which typically ranges from 0.7 to 0.9.

Finding 6: Soil Moisture. Numerous specialized soil moisture networks exist across the Nation, but there is no design for a national soil moisture network. The NASA Soil Moisture Active Passive Mission, planned for 2014, in conjunction with a robust *in situ* network for calibration, could provide adequate soil moisture data for many applications at regular intervals over much of the country if these efforts were implemented in a coordinated manner.

As the global air temperature has increased, one might expect an associated increase in pan evaporation, as higher temperatures provide more energy for evaporation. However, just the opposite has been observed (Roderick and Farquhar, 2002). There are at least three explanations for the observed decreases in pan evaporation: (1) higher humidity giving the atmosphere less capacity to take up water; (2) decreased winds (Vautard and others, 2010); and (3) reduced solar irradiance associated with increased cloud cover and aerosol concentration (Roderick and Farquhar, 2002). Reduced solar irradiance appears to be the best explanation for reduced pan evaporation in all settings around the world (Roderick and Farquhar, 2002). It has been hypothesized that increased cloud cover and aerosols concentrations will be a result of global temperature increases. Increased aerosol concentrations also are a result of atmospheric pollution.

Transpiration can be measured by using thermal sensors to measure sap flow in the stem of plants (Hatton and others, 1995). Transpiration measurements are made in experimental studies, but not routinely. The primary purpose for measuring transpiration is to separate transpiration from evaporation in measurements of ET, as described below.

There is a growing network of ET measurement sites. ET typically is measured using the eddy-covariance method (for example, Lee and others, 2004), which permits monitoring of carbon, water (that is, ET), and energy flux at a fast time step over the area of the flux tower footprint (about 1 k2). These measurements are increasingly recognized as an important tool for understanding the terrestrial water and carbon cycle. Nevertheless, "there are very limited direct measurements of actual evapotranspiration over global land areas, while global analysis products are sensitive to the type of analysis and can contain large errors, and thus are not suitable for trend analysis. Therefore, there is little literature on observed trends in evapotranspiration, whether actual or potential" (Intergovernmental Panel on Climate Change, 2007a).

Figure 9. Ameriflux sites in the United States at which ecosystem exchanges of water, energy, momentum, and CO_2 are measured continuously at short time intervals (Ameriflux, 2011).

The Ameriflux Network (Ameriflux, 21) consists of about active 90 sites (figure 9) that measure ecosystemexchanges of water, energy, momentum, and CO2 continuously at hourly or shorter time intervals. Ameriflux is part of the global Fluxnet network (http://daac.ornl.gov/FLUXNET/fluxnet.shtm) which consists of more than 500 flux towers. Information from sites in these networks is specific to the ecosystem (for example, vegetation type) in which the measurements are made.

Remote sensing provides perhaps the most useful approach for obtaining estimates of ET over large areas. Land-sat thermal, visible, and near-infrared data are being used to estimate ET at the field scale (Allen and others, 2007).These data can be used for scheduling crop irrigation, and also have been used to settle water-rights disputes. Further refinements can be made by using flux tower data in combination with remotely-sensed information (Anderson and others, 2007) or by using multiple sensors (Vinukollu, 2011).

Finding 7: Evapotranspiration. Operational evapotranspiration data can be obtained from remotely sensed information. *In situ* measurements of energy, carbon, and water fluxes above the canopy provide valuable information on water budgets (as well as energy and carbon budgets) and are valuable for improving remotely-sensed evapotranspiration estimates, but the number of such installations is relatively small.

Precipitation Amounts, Snowpack, and Glacier Mass

Precipitation is the primary means by which freshwater resources are replenished. Information about the magnitude and occurrence of precipitation is relevant to a wide range of decisions with significant economic implications, including long term infrastructure investments. Until recently water management and other resource dependent decisions have been made under the assumption that past observed precipitation patterns and magnitudes will continue in the future. The same assumption has been made about the occurrence of floods and droughts. There is evidence, however, that precipitation patterns are changing from those observed during the last 100 years, with some indication that storm intensity has increased (Zhang and others, 2007, Min and others, 2011), although this evidence is not conclusive (Villarini and others, 2011).

There are many uses for precipitation data. Watershed-models based on historical rainfall records are routinely applied to estimate flood frequency at ungaged sites, or sites with short periods of streamflow record. These flood frequency estimates are used to evaluate and design bridges, flood control structures, stormwater management facilities, and other water-related infrastructure in flood-prone areas, as well as to produce Flood Insurance Rate Maps. Data on daily rainfall and interannual variation in precipitation are important to support agricultural operations. Rainfall data are used in National Weather Service flash flood, flood, and streamflow forecasts.

The NWS precipitation frequency estimates (Technical Paper 24, U.S. Weather Bureau, 1953), updated in the early 1960s, have become the *de facto* national standards by virtue of their inclusion or reference in widely used design standards. NOAA's Hydrometeorological Design Studies Center (HDSC) (National Oceanic and Atmospheric Administration, 2011b) began an effort in 2003 to update precipitation frequency estimates, and updated results are available for 19 States. Probable maximum precipitation estimates were developed decades ago (National Oceanic and Atmospheric Administration, 2011c), and NOAA has no plans to update these estimates. Advances in science, monitoring, and information technology provide an opportunity to revisit these estimates and incorporate newer data. A systematic and consistent approach that uses recent data is needed for updating precipitation frequency estimates. Development of a sound basis for a conservative design standard is important in addressing the impacts of a changed and changing climate.

Finding 8: Precipitation Frequency-Duration- Intensity Estimates. A systematic and consistent approach that includes recent data and improved statistical techniques is needed for updating precipitation intensity-duration- frequency estimates.

NOAA manages the Climate Reference Network (114 stations) and the U.S. Historical Climatology Network (1,221 stations) (National Oceanic and Atmospheric Administration, 2011d) networks, which are designed to document temperature and precipitation trends. Of the currently operating 11,400 Cooperative Observer Program (COOP) (National Oceanic and Atmospheric Administration, 2011e) stations, about 5,000 make up the "climate" network (the remaining stations support hydrology requirements). In general, precipitation gages tend to be more numerous in the east, south central, southeast, southwest, and northwest, often co-located with streamgages or installed at or near major metropolitan airports. Although airports are convenient, the continuity and accuracy of the long-term data sets can be inconsistent due to site relocations, instrument changes, and the consequences of encroaching urbanization and related heat island effects. With a few exceptions, the instruments used by NWS cooperative observers have not changed significantly over the past century. Numerous other agencies and group measure precipitation. Included are the USGS (for example, http://waterdata.usgs.gov/va/nwis/current/?type=precip&group_key=county_cd), numerous state networks (for example, http://www.nc-climate.ncsu.edu/map/), research institutions, and private individuals (for example, Weather Stations at http://www.wunderground.com/cgi-bin/findweather/hdfForecast?query=20192).

The NWS NEXRAD system (National Oceanic and Atmospheric Administration, 2011f) revolutionized precipitation reporting on the U.S. NEXRAD provides real-time information in precipitation intensity and other measures. Storm precipitation amounts can be obtained by integrating individual NEXRAD scenes. Data are available at a spatial resolution of 4.4 x 4.4 km, and can be statistically down-scaled to 1.1 x 1.1 km. Radar estimates of precipitation, however, must be regularly calibrated to observations. Calibration schemes account for such variables as season, precipitation type, pre- cipitation intensity, and weather patterns (Li and others, 2008).

Precipitation also can be estimated from satellites, and with increasing resolution and accuracy. Satellite precipitation estimates are less accurate than gage and radar measurements, but satellite estimates can be provided in near-real time (an advantage over many gages), and everywhere, including over the oceans and in mountainous regions (an advantage over both gages and radar). Numerous satellite precipitation products are available including STAR (National Oceanic and Atmospheric Administration, 2011g; CMORPH (National Oceanic and Atmospheric Administration, 2011h); TMPA-RT, which is produced byTRMM ((National Aeronautic and Space Administration, 2011a); and PERSIANN (CHRS-PERSIANN- CCS, 2011). As with radar, these estimates require some calibration and bias correction, especially for hydrologic applications (Behrangi and others, 2011; Tang and others, 2010). As with *in situ* gages, satellites require maintenance and eventual replacement.

Finding 9: *In Situ* Precipitation Measurement. There is an opportunity to form a robust precipitation measurement network in the U.S. with existing gages operated by numerous Federal, State, and local agencies, research organizations, and private citizens. However, lack of coordination, protocol, and communication among the organizations impedes formation of such a network. Because of this lack of coordination among precipitation gage operators, there has been no broad analysis of precipitation measurement needs.

The NRCS has been collecting snow and climate data and producing water supply forecasts in 12 western States and Alaska for 76 years (U.S. Department of Agriculture, 2011b). As part of this collection system, many of the automated SNOTEL (U.S. Department of Agriculture, 2011e) sites, which transmit data in near-real time, have been in operation since the middle 1970s and presently the network consists of 813 stations. NRCS and its many partners also collect manual snow course data each month during the snow season at up to 900 sites throughout the Western U.S. and Alaska. NRCS has proposed to coordinate and provide technical leadership for collection and dissemination of snow information over a greater area of the U.S. (National Resources Conservation Service, 2011), which could enhance products from NOAA's Snow Data Assimilation System, described below.

Finding 10: Remotely-Sensed Precipitation Measurements. Radar and satellite precipitation measurement capabilities have provided precipitation estimates over areas not covered by ground-based gages. These estimates, however, rely on gage data for calibration and bias correction.

NOAA's National Operational Hydrologic Remote Sensing Center NOHRSC (National Oceanic and Atmospheric Administration, 2011i) integrates airborne and satellite snow observations from all available electronic sources for the coterminous U.S. These data are used along with data from SNOTEL (U.S. Department of Agriculture, 2011b) to generate estimates of snowpack characteristics generated by a physically-based snow model to generate the National Snow Analyses (NSA) (U.S. Department of Agriculture, 2011c) on a daily basis for the coterminous U.S., as shown in figure 10. Ground-based and remotely-sensed snow observations are assimilated daily into the simulated snow-model state variables including snow water equivalent, snow depth, surface and profile snowpack temperatures, snowmelt, surface and blowing snow sublimation, snow-surface energy exchanges.

The SNOw Data Assimilation System (SNODAS) is a modeling and data assimilation system developed by NOHRSC to provide the best possible estimates of snow cover and associated parameters to support hydrologic modeling and analysis.

The aim of SNODAS is to provide a physically consistent framework to integrate snow data from satellite, airborne platforms, and ground stations with model estimates of snow cover. SNODAS is an important resource but it currently lacks a reanalysis of its data.

Without this, a climatology cannot be developed, so there is no way to put the current SNODAS measurements into a historical context. Development of such a climatology would be an important step for the product.

Finding 11: Snowpack Monitoring. The Natural Resources Conservation Service's SNOTEL (SNOwpack TELemetry) network operates in 12 western States and Alaska. NRCS has proposed to coordinate and provide technical leadership for collection and dissemination of snow information over a greater area of the U.S., which could enhance products from NOAA's Snow Data Assimilation System.

Almost all of the world's freshwater is located in glaciers and icecaps, and concerns about the effect of glacier loss on water supplies (although not necessarily in the U.S.), sea level [potentially tens of meters of sea-level rise (U.S. Geological Survey, 2000)], and ocean circulation support continued monitoring and research. Glaciers in the U.S. are monitored by the National Park Service (Denali National Park, North Cascades National Park, Glacier National Park, Mount Ranier National Park, and elsewhere; for example, North Cascades National Park Complex at http://www.nps.gov/noca/naturescience/glacial-mass-balance1.htm), the USGS (three benchmark glaciers; for example, USGS Benchmark Glaciers at http:// ak.water.usgs.gov/glaciology/), and academic researchers. The World Glacier Monitoring Service (http://www.geo.uzh.ch/ microsite/wgms/) compiles information from global glacier monitoring, and the Service provides guidelines and standards for glacier monitoring.

Much glacier monitoring can be conducted from satellites (for example, Landsat, ASTER, and SPOT), but ground-based measurements are required for calibration and testing of remote-sensing methods (for example, Paul and others, 2007).

Finding 12: Glacier Monitoring. Glacier monitoring is important, given the potentially large effect glacier meltwater could have on sea level and ocean circulation. A combination of ground-based and satellite monitoring is warranted. As in many monitoring efforts, there is some lack of coordination among existing monitoring activities.

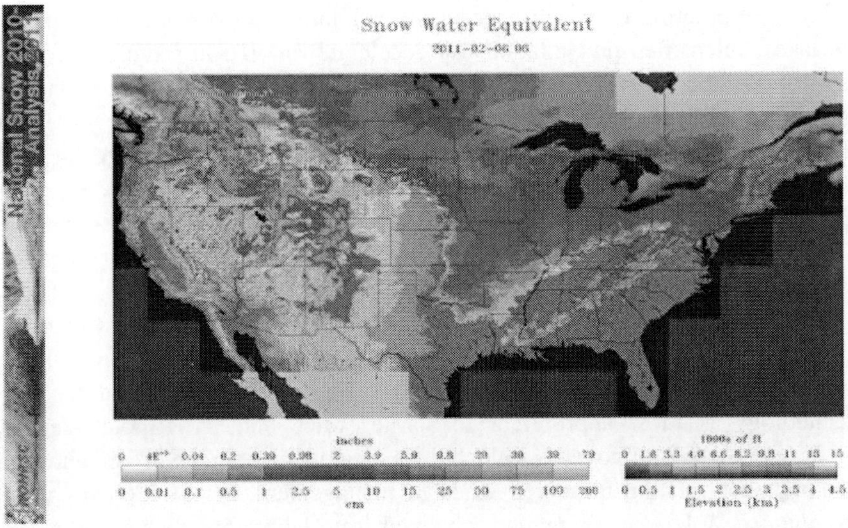

Figure 10. Example map of snow-water equivalent for the coterminus U.S. during February 2011 (U.S. Department of Agriculture, 2011a). Animated maps can be produced to display daily variations in snow depth, snowpack temperatures, sublimation, and other statistics.

Water Quality and Ecosystem Condition

Water-quality data include physical characteristics such as temperature and suspended solids; chemical characteristics such as pH, alkalinity, minerals, metals, nutrients (nitrogen and phosphorus), organics (including naturally occurring substances; pesticides, pharmaceuticals, and other anthropogenic compounds; and byproducts or degradation products such as nitrosamines or other disinfection byproducts); and microbial characteristics such as viruses, bacteria, protozoa, and algae. Many of these parameters are interrelated and have direct public health implications (Noyes and others, 2009).

Water-quality data are measured in a variety of media, including surface waters, groundwater, precipitation, sediment, and plant and animal tissue. Hydrologic conditions under which data were collected are a key part of the data set. Without this information, water-quality data are less valuable and more difficult to interpret.

Climate influences water quality directly through changes in temperature and precipitation patterns. Changes in temperature also affect microbial and chemical reaction rates including those controlling degradation of contaminants, solubility, and pathogen survival and die-off. Temperature affects evaporation rates and consequently salinity (salt content), particularly in lakes. Intensified precipitation can accelerate runoff and sediment erosion and deposition rates, as well as atmospheric deposition into water bodies. Drought and heavy precipitation events have been correlated with increased incidence of waterborne disease (Nichols and others, 2009). Changes in land and water management will affect contaminant inputs to freshwater, both positively and negatively. The effects of these types of changes can be immediate and profound (Scanlon and others, 2007), with climate change potentially adding to the effects of these other stressors.

Some types of water-quality data can be collected using *in situ* sensors to provide real-time data. Remote sensing and satellite imagery can provide spatially-coarse information on algal blooms, turbidity, and temperature. This technology is most appropriate for large water bodies. Remotely-sensed information on land cover is valuable for evaluating sources of contaminant loadings for relating measured loads to management actions. Perhaps most commonly, water-quality data are obtained from laboratory analyses of water samples collected at specific locations (streams, groundwater, wastewater, drinking water systems, outfalls, intakes, etc.). The specific types of water-quality data that are collected and the frequency of sampling depend on resource availability, sampling protocols, and regulatory requirements.

Water-quality data are collected by Federal, State, Tribal, and local agencies, interstate commissions, the academic community, and citizen volunteers for a variety of purposes. There have been numerous reports in recent years indicating that this level of monitoring is insufficient and/or not well enough coordinated to provide comprehensive information about U.S. water resources (for example, U.S. Government Accountability Office, 2004; H. John Heinz Center for Science, Economics, and the Environment, 2002). This is due, at least in part, to an inability to access and integrate the data that are collected by the various organizations engaged in water-quality monitoring.

The USGS collects and analyzes chemical, physical, and biological characteristics of water, sediment, and tissue samples from across the Nation and compiles the data into the NWIS database (U.S. Geological Survey, 2011b). The USGS National Stream Quality Accounting Network (NASQAN) Network (U.S. Geological Survey, 2011g) consists of 21 sites on major drainages to the Mississippi River, as well as the Mississippi River itself (Figure 11). The network originally supported about 618 sites in the mid-1970s (Alexander and others, 1997). This decline in the number of sites reflects the challenges of maintaining a robust monitoring network with level funding.) The USGS National Water-quality Assessment Program (NAWQA) collects samples at 113 sites, some of which are sampled annually, some every two years, and some every 4 years (Figure 12). An additional 58 wadeable streams are sampled for water chemistry and ecological conditions. Additionally, about 1,300 surface-water and groundwater sites are instrumented to continuously record physical and chemical characteristics of the water including pH, specific conductance, temperature, dissolved oxygen, and percent dissolved-oxygen saturation (U.S. Geological Survey, 2011a). Investments in sensor technology would greatly enhance the Nation's water-quality monitoring capabilities.

The USEPA conducts sampling designed to statistically answer questions about the overall quality of the Nation's waters. The Wadeable Streams Assessment (U.S. Environmental Protection Agency, 2006) sampled 1,392 randomly selected sites. Macroinvertebrates were used to assess the condition of the streams relative to reference (unimpacted) streams. Following a similar design as the Wadeable Streams Assessment, the National Lakes Assessment sampled 1,028 lakes in the U.S. in 2007 in order to obtain a statistical measure of the health of the Nation's lakes (U.S. Environmental Protection Agency, 2010a). These assessments are designed to be repeated at regular intervals.

States conduct both statistical surveys and targeted monitoring under USEPA guidance. States are required to report to USEPA every two years on the condition of their surface waters, as required under sections 303(d) and 305(b) of the Clean Water Act. The reports generally identify the water bodies that are impaired for the designated uses and the causes of impairment. USEPA maintains a database of the State reports (U.S. Environmental Protection Agency, 2011c).

A National Water-Quality Monitoring Network design was developed by a large group of participants through the National Water-Quality Monitoring Council (2008). The network design is consistent with the National Ocean Council's Strategic Action Plan (2011) priority objectives on Water Quality and Sustainable Practices on Land (objective 7) and Ocean, Coastal, and Great Lakes Observations, Mapping, and Infrastructure (objective 9). If implemented, the network will provide integrated and coordinated data on the Nation's water quality from the uplands to the coasts. The design is based on five objectives to answer important management questions (National Water-Quality Monitoring Council, 2008). The network also is designed to link with the Integrated Ocean Observing System; is based on a multi-disciplinary approach that integrates water resources from the uplands to the coast, integrating physical, chemical, and biological measures; and stresses the importance of metadata, comparable methods, quality assurance, and data archival. The rivers component of the network (Figure 12) includes 258 sites representing 90 percent or more of the flow from each Hydrologic Unit Code sixth- level (HUC-6) watershed (U.S. Geological Survey, 2011h) in the country, 72 sites at locations draining directly to key estuaries, and 56 sites to document flows and loads to the Great Lakes. Currently, 5 stations of the network are funded by the USGS as a sub-network of the National Stream Quality Accounting Network (U.S. Geological Survey, 2011i).

Finding 13: Water-Quality Networks. Extensive water-quality monitoring is conducted in the U.S., and various networks have evolved to meet user needs. Integration of data from the various networks could offer benefits, if data were comparable and well integrated. The proposed National Monitoring Network offers a sound basis for understanding fluxes of key contaminants from major basins across the Nation.

Water-quality issues associated with climate change are likely to have multiple public health impacts due to shifts in the microbial ecology of freshwater and coastal systems, new and different pathogens and chemical contaminants, and an increase in algal blooms.

Heavy precipitation events have been statistically linked to an increased incidence of waterborne disease outbreaks in the United States (Interagency Working Group on Climate Change and Health, 2010a; Curriero and others, 2001). The anticipated increase in the risk for waterborne disease with climate change will require improved surveillance systems and coordination of prevention efforts at local, State, and national levels. While the types of data systems needed to monitor the incidence and prevalence of waterborne disease differ from other monitoring networks, it is important to integrate geospatial water resource data with public health data. The Centers for Disease Control and Prevention's (CDC) National Environmental Public Health Tracking Network (http://ephtracking.cdc.gov/showAbout.action) could be a potentially valuable source of information for tracking waterborne disease data and for protecting human health.

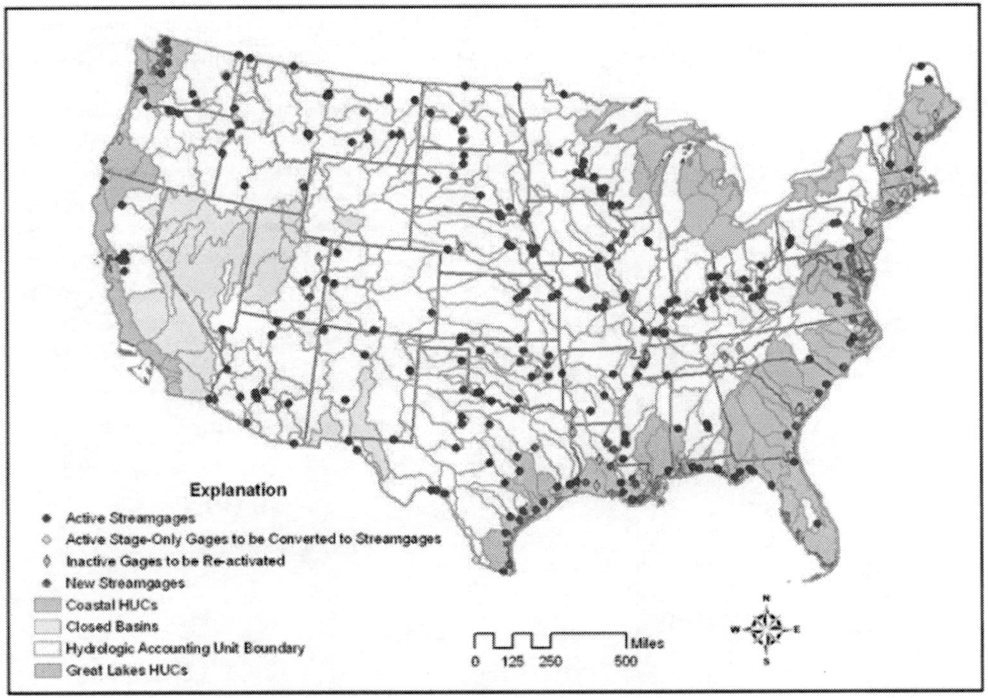

Figure 11. Map of the conterminous United States, showing Hydrologic Accounting Unit boundaries, closed basins, coastal basins, Great Lakes basins, and proposed sites for the National Water-Quality Monitoring Network (National Water-Quality Monitoring Council, 2008).

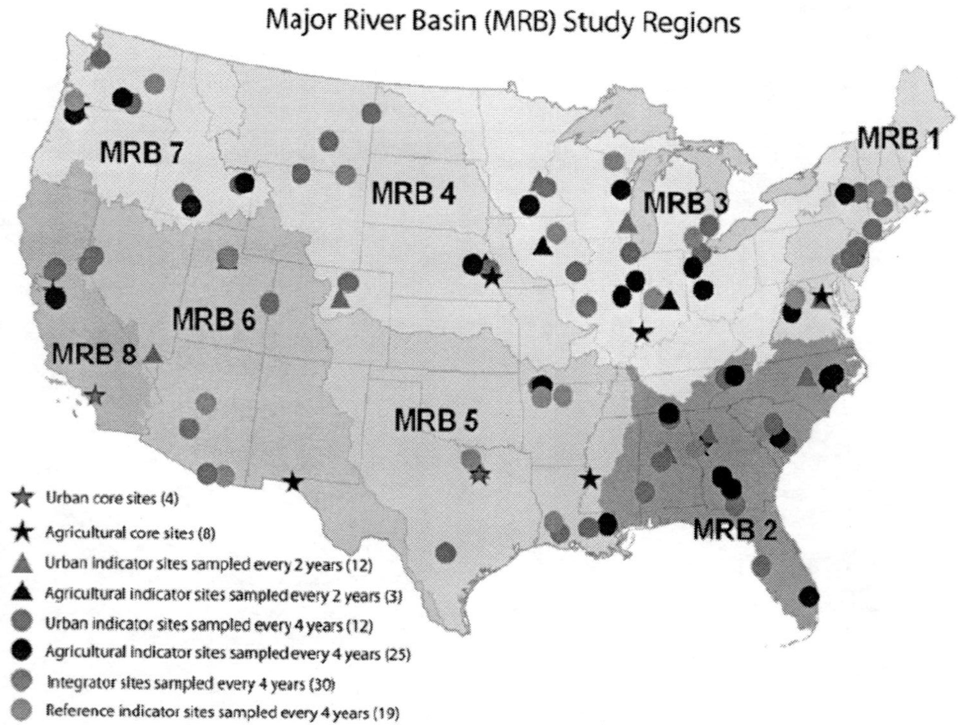

Figure 12. U.S. Geological Survey National Water-quality Assessment Program water-quality trend sites (http://water.usgs.gov/nawqa/studies/mrb/images/mrb_sites_lg.png).

Key issues include:

- Endemic disease and outbreaks associated with exposure to waterborne pathogens;
- Surveillance of diseases associated with waterborne contaminant exposures (chemicals and their byproducts and metabolites, toxins released by algae and cyanobacteria, etc); and
- Surveillance of developmental health effects relevant to waterborne contaminants (microbial and chemical).

Finding 14: Waterborne Pathogens. Prevalence of waterborne pathogens, including harmful algal blooms, likely will change in response to rising temperatures, as well as to alterations in flow characteristics and contaminant loadings. Systems for tracking these changes are not well developed, nor are they well integrated with geospatial, water-quality, and public health information.

Wetlands are sensitive to changes in hydrologic conditions. Small changes in groundwater levels can determine whether a wetland thrives or disappears. The effect of climate change on wetlands will depend on the location of the wetland within the hydrologic landscape, and the magnitude of temperature and precipitation changes (Winter, 2007). Changing sea level already is impacting coastal wetlands, causing wetland retreat in many locations (for example, Williams and others, 1999; Lanthrop and Love, 2007). USEPA has begun a National Wetland Assessment (http://water.epa.gov/type/wetlands/assessment/survey/index.cfm#about) designed to provide regional and national estimates of wetland ecological integrity. The assessment will use a probability-based design, similar to that used for the Wadeable Streams Assessment, to produce statistically-valid estimates of conditions for various wetland types.

The Wetlands Mapping Standard was developed and formally endorsed by the Federal Geographic Data Committee in response to the need for a higher degree of wetlands data. However, implementation of the standard by States is not yet fully realized. Use of the standard will help provide a more complete and accurate picture of wetland resources in the United States. Wetlands are inextricably tied to water levels and changes in climatic conditions affecting water availability will greatly influence the nature and function of specific wetlands, including the type of plant and animal species within them. In developing this mapping standard, an effort was made to identify and accommodate technology and map-scale enhancements that will ensure long-term use of the standard and minimize the need for revisions and updates.

Finding 15: Wetlands. Freshwater and coastal wetlands are sensitive to changes in hydrologic and sea-level conditions. The National Wetland Assessment should provide valuable information on the conditions of a sampling of the Nation's wetlands, and a basis for future comparisons, although information on specific systems will not be available Implementation of the new wetlands mapping standard needs to be expedited.

Water Withdrawals and Consumptive Use

Water withdrawal data generally are collected by States, water management districts, and utilities. In many cases, withdrawals are estimated from proxies, rather than directly. Information is collected and reported differently from State to State, and almost no data are available on consumptive use (as opposed to total withdrawals of water) (National Research Council, 2002).

The USGS works in cooperation with local, State, and Federal environmental agencies to compile, and disseminate water withdrawal information through the National Water-Use Information Program (NWUIP) (U.S. Geological Survey, 2009a) at five-year intervals. Information on consumptive use has not been reported since 1995. Data are searchable through the USGS National Atlas (U.S. Geological Survey, 2011j). The USDA conducts an agricultural census (U.S. Department of Agriculture, 2011d) and also conducts a farm and ranch survey (FRS) to provide a snapshot of agricultural water use (U.S. Department of Agriculture, 2008). These data are compiled and summarized on the basis of political boundaries (States, counties) and do not necessarily reflect watershed characteristics.

Numerous reports have documented the close link between energy production and water use (see Mielke and others, 2010, for a recent review). Water is required for energy extraction, transportation, processing, and production. Energy is used to withdraw, treat, and transport water. The Department of Energy collects some water withdrawal data through the Energy Information Administration (U.S. Energy Information Administration, 2011); a recent GAO (U.S. Government Accountability Office, 2009) report has recommended some enhancements to that program. Changes in the Nation's energy portfolio (increased use of biofuels; new hydrocarbon energy extraction methods) and increased use of carbon sequestration, already are having an effect on water resources. Monitoring networks do not exist to address these issues (U.S. Government Accountability Office, 2009).

The national Water Census (U.S. Geological Survey, 2007) and the Department of Interior's (DOI) WaterSMART Program (U.S. Department of the Interior, 2011) are important steps towards providing a comprehensive perspective on water resources. The Water Census, which is modeled in large part on 2002 recommendations from the National Research Council (National Research Council, 2002) is designed to describe the changing availability, quality, location, and uses of water resources. This key information needs to be readily available to water managers and the public.

The Water Census will be organized on a watershed basis and contain:

- Water use information: withdrawals, return flows, and consumptive use for all water use sectors with special data collection efforts designed to track changing water use patterns relevant to public water systems, agriculture, energy, industry, and non-potable demands;
- Water quantity and availability information including analysis of changes in storage (groundwater) and flow (rivers);
- Analysis of changes in water quality that are critical to its use (temperature, sediment, nutrients, pathogens, salinity, waterborne contaminants) and;
- Identification of alternative water sources that may become available through technological advances including reuse, green infrastructure, conservation, desalination, aquifer storage and recovery and evolving energy production and industrial water use practices.

As a part of the Water Census design, there are plans to provide grants to States to help improve the quality of the Nation's withdrawal and consumptive use data, and to develop common methods for collecting and reporting the data.

The concept of water use is expanding to include environmental flows, or the flows required to maintain a functioning and healthy ecosystem, such that rivers and streams are now considered important 'users' of water (Naiman and others, 2002; Postel and Richter, 2003). The concept embodies the provision of suitable habitat for living aquatic resources—both plants and animals. Alterations of flow from natural conditions are known to elicit an ecological response (Poff and Zimmerman, 2010). Flow alterations are defined as any change from natural conditions in the magnitude, frequency, duration, timing, and rate-of-change of flows.

Finding 16: Withdrawals and Consumptive Use of Water. Water withdrawal and consumptive use data, particularly as related to the Nation's energy portfolio, are inadequate for making long-term decisions about water availability in the U.S.

The management of freshwater systems to meet both human and ecological needs remains a major challenge (Arthington and others, 2006; Poff and Zimmerman, 2010). For example, in the Pacific Northwest, projects to improve habitat for salmon are "moving forward with little or no knowledge of specific linkages between restoration actions and the responses of target species." (National Oceanic and Atmospheric Administration, 2011j). Stream-by-stream and species-by-species analysis of environmental flow needs is not practical. Even when environmental flow needs are well defined, the challenge of maintaining a balance between environmental flows and human needs for water supply, waste assimilation, navigation, power generation, and other uses can be difficult and contentious. New data and research on the environmental flow requirements, including pre- and post- alteration information, are needed for the restoration and maintenance of healthy aquatic ecosystems.

Finding 17: Environmental Flows. New data and research on the environmental flow requirements, including pre- and post- alteration information, are needed for the restoration and maintenance of healthy aquatic ecosystems.

Review Element 3: Improve Data Management to Increase Efficiency of Data Acquisition and Reporting

Review Element from Section 9506:

"to establish data management and communication protocols and standards to increase the quality and efficiency by which each Federal agency acquires and reports relevant data..."

Robust tools for managing data are essential for effective decision-making. Data management practices have advanced over the past decade due to the evolution of information technology, improved communication tools, and improved decision support tools.

Currently there are dozens of data management approaches (see Appendix B for some important examples) that are in use and a host of databases that house various types of water data and information The interoperability of these systems varies and there are ongoing efforts to improve coordination. Obstacles to data sharing include differing geospatial reference systems, lack of common definitions for the same measure, and inadequate metadata. Most data are accessible via the internet and digital copies of the data can be retrieved.

A major challenge facing producers of water information is ensuring that data are accessible, quality-controlled, and available in formats that facilitate their use by stakeholders. Over the past decade, a number of Federal agencies and other institutions have expanded their water databases, and developed a range of information management tools to help ensure compatibility and interoperability of disparate data sets so they can be applied to a variety of purposes. The USEPA maintains a water-quality data warehouse called STORET (U.S. Environmental Protection Agency, 2011d) that catalogues water-quality data by location and properties (physical, chemical, biological and habitat). STORET is updated regularly and its data record ranges from the 1970s to the present. Data in STORET tend to be focused on water or wastewater treatment facilities or watersheds of concern. Because the data are derived from different entities, all inputs must document a standardized set of metadata that include: the sampling and analytical methods used; in appropriate situations, the laboratory used to analyze the samples; and the quality control checks used when sampling, handling the samples, and analyzing the data.

Finding 18: Interoperable Data Systems. Ready access to the full range of hydroclimatic data collected by government agencies and other interests is inadequate. Data are collected using a range of protocols, which are not always documented, and are archived in a variety of ways, from modern relational databases to paper copies in files. There is much to be gained from use of consistent documentation standards and improvements in interoperability of data systems.

While NWIS and STORET are designed for different purposes, the systems are developing collaborative websites to provide data to users in a common format. These systems have developed standardized metadata, compatible search parameters and common vocabulary that will allow for greater functionality of available data to be used for climate change analysis.

This approach will allow data users to access data from both databases. Efforts are underway to expand this effort to include other data systems, thus allowing for greater consistency across a wider range of water body types.

A variety of institutions are exploring options for enhancing interoperability. One effort that has seen some initial promising results is the Consortium of Universities for the Advancement of Hydrologic Science, Inc. (CUAHSI), Hydrologic Information System (HIS) project, funded by the National Science Foundation (NSF) and led by a multi-institutional research team. This group has been studying how to synthesize the water data by means of a services-oriented architecture. A large scale prototype water information net- work has been constructed that presently provides access to over 50 water-data services including significant proportions of the water data holdings of the USGS and USEPA, information from NWS, National Climate Data Center (NCDC), USDA-NRCS and other agencies at the Federal and State level, as well as water data from a dozen universities (see http://hydrodesktop.codeplex.com/). HIS is used by multiple agencies in the State of Texas to share data operationally and is being promoted by the Niagara Peninsula Conservation Authority to link information across cities along the Great Lakes. CUAHSI has defined a language called WaterML (Water Markup Language) for conveying time series of water observations data through the internet. The USGS and some other water agencies now publish some of their observations data in WaterML. USEPA has developed a web services language called WQX (Water Quality Exchange) that conveys groups of water-quality observations, and CUAHSI has developed a translator that converts WQX into WaterML so that time series of physical hydrology and water-quality data can be acquired in a consistent way.

The Open Geospatial Consortium (OGC) (http://www.opengeospatial.org/) is an international organization, representing about 400 companies and agencies, which has developed the most widely used standards for sharing geospatial data through the internet. In 2008, CUAHSI proposed to the OGC that there should be established a Hydrology Domain Working Group to harmonize WaterML with OGC standards, and later the OGC and the World Meteorological Organization expanded this mission to included joint development of data standards for hydrology, climatology, oceanography and meteorology.

The CUAHSI Hydrologic Information System project has concluded that the best approach to enhance internet-based water data sharing in the United States is to adapt existing OGC data standards. Climate and weather grid information can similarly be searched and accessed by being indexed in a consistent way with water resources time series. The ACWI/FGDC Subcommittee on Spatial Water Data supports this approach to developing water data services in the United States. Leadership at the Federal level will be needed to develop strategies to enhance water data sharing across all levels of government, among water disciplines, and between water science and water management in the United States and with bordering nations.

The U.S. Integrated Ocean Observing System (IOOS®) is a national observing system that integrates, coordinates and enhances efforts to deliver coastal and near-coastal information to decision-makers to improve safety, enhance our economy, and protect our environment. The integration of information facilitated by open source standards and protocols, such as the Open Geospatial Consortium Sensor Observation Service, and data formats that enable more efficient, routine, and effective use of data is key to building IOOS®. IOOS® and the National Water-Quality Monitoring Network for U.S. coastal waters and their tributaries have been coordinating to bridge water-quality data collected in rivers and watersheds with data from the estuaries to coastal waters and the Great Lakes in order to facilitate seamless use of the data.

A reliable system to monitor and report the occurrence and prevalence of waterborne disease would aid in understanding the role of climate and other driving factors in the distribution of those diseases. Officials at the Centers for Disease Control and Prevention (CDC) have recognized the need for such a system and it is important that this effort is coordinated with development of data systems focused on changes in the natural environment.

In general, hydroclimatic data are insufficiently integrated (or integratable by the user community) to support important management decisions. For example, most water-quality data are not integrated with water quantity data. Likewise, groundwater and streamflow data generally are not integrated, thereby limiting, for example, understanding the effects of groundwater pumping on streamflow. Hydroclimatic data are not connected to issues of social relevance. For example, the capacity does not exist to link information about vulnerable human populations with data about implications of changes in the hydroclimatic cycle (that is, frequency and magnitude of floods and droughts).

Accelerated progress is needed on:

- Multi-agency cooperation to deliver information seamlessly;
- National integration of observational data and meta-data;

- Adoption of common operational standards, quality assurance procedures, and data exchange formats;
- Data interoperability;
- Coordination of all types of hydroclimatic monitoring; and
- Linkage of hydroclimatic data to issues of social relevance.

These efforts need to be coordinated with hydrogeological and biogeochemical modeling of freshwater resources and ensure the right types of data are being collected at the right locations and frequency.

Finding 19: Data and Decision-Making. In general, hydroclimatic data are insufficiently integrated (or readily integratable by the user community) to support important management decisions, and hydroclimatic data are inadequately connected to information on issues of social relevance.

Review Element 4: Options for Establishment of a Data Portal

Review Element from Section 9506:

"to consider options for the establishment of a data portal to enhance access to water resource data..."

There are a number of ongoing efforts within single agencies or jointly among agencies to develop data portals that allow for data integration. Easy internet access to hydroclimatic information and data is a high priority for all stakeholders. As an example, the USGCRP has proposed establishing an "Interagency Climate Information Portal" to provide web-based access to critical climate information, including climate data, models, resources and tools, for decision makers and the public.

As data are coordinated, it is clear that the quantity of data available for a given watershed depends on basin location relative to issues that drive monitoring and resources available for consistent data collection. There are opportunities for integrating geospatial analyses with sensor data through a portal.

An important challenge is to ensure that a data portal is designed to maximize access to and utility of the data. As a result, decision-makers and other end-users need to play an integral role in portal development and testing, and the entry point needs to be self-explanatory and allow for data to be accessed by a range of users with different levels of experience and expertise. Language, semantics, data definitions, and criteria need to be developed collaboratively with inputs from all stakeholders (decision-makers, modelers, information technology, database developers, the user community, etc.) so that information is accessible and useful to the appropriate audiences and guidance is provided on best management practices. Portal design needs to consider how to best link information to its original source to enable users to query the data in different ways and access more details as necessary. The portal design needs to be flexible enough to adapt to changing information technology. Robust quality standards are needed to ensure the credibility and reliability of the portal.

Finding 20: New Data Portals. A new effort by the U.S. Global Change Research Program, using existing agency resources, has been initiated to build an interagency data portal, and this portal also may provide access to hydroclimatic data. The initial phase of this process, conducted with strong coordination among Federal and non-Federal partners, is expected to focus on data generated in support of the National Climate Assessment. The portal also is expected to provide guidance on appropriate use of data and on the use of probabilistic information. Other efforts to better organize data, such as the Integrated Water Resources Sciences and Services activity, also should contribute to enhanced data access.

Review Element 5: Adequacy of Hydrologic and Other Models

Review Element from Section 9506:

"to facilitate the development of hydrologic and other models to integrate data that reflects ground-water and surface water interactions…"

Successful management of our Nation's water resources requires an integrated approach. Models provide decision-makers with tools that can be used to simulate and visualize water quantity and quality under the range of climatic, hydrologic, and environmental conditions that reflect the current and future state of water resources. Models can be used to help explore potential future climate scenarios at relevant scales for:

- Prediction of basin response to extreme rainfall and snowmelt through application of precipitation-runoff models to evaluate changes in water-balance relationships, flow regimes, flood peaks and volumes, soil-water relationships, sediment yields, and groundwater recharge resulting from climate change;
- Evaluation of the effect of predicted climate scenarios on streamflow, sediment delivery, and water balance in watersheds;
- Evaluation of interactions between streamflow and groundwater levels; and
- Planning and design of urban drainage systems, flood control structures, water-supply storage, and other water infrastructure.

Models can generally be classified as statistical or process-based, although process-based models can be applied statistically. Process-based models are those based on underlying physical, biogeochemical, and ecological processes.

As previously noted, statistical hydrologic models have been based on the assumption that past distributions of precipitation and basin response are a good indicator of future distributions. It generally is recognized now, however, that this assumption of stationarity is not valid (Milly and others, 2008), and likely never was valid. As a result, statistical understanding and models of future precipitation and stream-flow characteristics needs to be improved. Models describing frequency of large precipitation events and the magnitude and intensity of those events also need to be improved. Likewise, statistical models of flood and drought recurrence and magnitude must be modernized in order to effectively plan for future conditions.

The capacity of process-based models is continuing to advance improvements in computing power, visualization capabilities, incorporation of remotely-sensed data, and research leading to better understanding of controlling processes. Integration of processes (for example, groundwater and surface water; hydrology and water chemistry) and feedbacks among processes (for example, hydraulic conditions and channel morphology; land-climate interactions) is a challenge, although progress is being made (Markstrom and others, 2008).

While there are significant efforts underway to develop ecosystem models and evaluate the services that ecosystems provide, there is a need to better integrate work on ecological processes with water resources data, models, and tools. Models for predicting how aquatic habitats respond to changing hydrology, water quality, and climate are needed. As aquatic habitats change, plant and animal communities will also adapt to changes in flow and water quality and a better understanding of these interrelationships is needed to evaluate the potential outcomes of climate mitigation and adaptation approaches. As a result, it is difficult to forecast the full range of effects of management actions and/or climate change on the entire resource.

Finding 21: Hydroclimatic Statistics and a Changing Climate. Statistical models of hydroclimatic events that are based on the assumption that future conditions will have the same statistical properties as past conditions are not valid. Improved understanding of the statistical characteristics of precipitation and streamflow are needed for effective planning and management under uncertainty.

Regional climate models or down-scaled general circulation models/global climate models (GCMs) also are evolving and helping to provide insight on possible future climate scenarios. It is fairly clear, however, that current down-scaled GCM's do a poor job in hindcasting past precipitation and hydrologic conditions when climate is known (for example, Stephens, and others, 2010; Bae and others, 2011), so confidence in future predictions of hydrology is weak. As a result, there is a need to (1) more rigorously test down-scaled GCM's for water-resources planning applications, (2) develop guidance for appropriate applications, (3) document uncertainty in results, and (4) understand the implication of new finer-scaled GCMs for hydrologic applications.

Community modeling is gaining increasing acceptance as an alternative to proprietary or closed-source models. Community models serve the diverse needs of a particular modeling community and model capabilities are advanced through contributions from the entire community. As an example, NOAA's Community Hydrologic Prediction System (CHPS) (National Oceanic and Atmospheric Administration, 2011k) provides an open architecture that integrates observations, analytical tools, and models to produce water resource forecasts for a wide array of users (Figure 13). Approaches also are under development to downscale GCM outputs to regions and localities, to assist in resource decision-making (for example, Means and others, 2010). Some of the key objectives of current modeling efforts include understanding land surface-atmospheric interactions, complex terrain, and the role of decadal scale oscillations in ocean temperatures.

Finding 22: Integrated Modeling. Although there have been recent advances, current coupled groundwater-surface water models, and fully coupled hydrologic-water quality-ecosystem models are inadequate for fore- casting the full range of effects of management actions and/or climate change on the entire resource.

The availability of reliable, up-to-date, readily accessible data is crucial to the effective development and use of models. Important data include:

- Components described in **Review Element 1**: Stream-flow and surface-water data, groundwater-level data, soil moisture and evapotranspiration, precipitation (including rainfall, snowfall, snowpack characteristics, soil characteristics, and glacier mass), water quality, and water use information;
- Climatic measures such as temperature, wind speed, and direction;
- Physical measures of the landscape such as elevation, slope, and land use.
- Land cover;
- Societal demographics; and
- Paleoclimatic data for understanding the severity of past droughts and flood.

Finding 23: Global Climate Models. There is a need to develop and more rigorously test down-scaled Global Climate Models (GCM's) to provide more reliable projections of water-resources conditions for planning applications, and develop guidance for appropriate applications.

Review Element 6: Apply Hydrologic Models to Water Resource Management Problems

Review Element from Section 9506:

"to apply the hydrologic and other models developed under paragraph (5) to water resource management problems identified by the panel, including the need to maintain or improve ecological resiliency at watershed and aquifer system scales..."

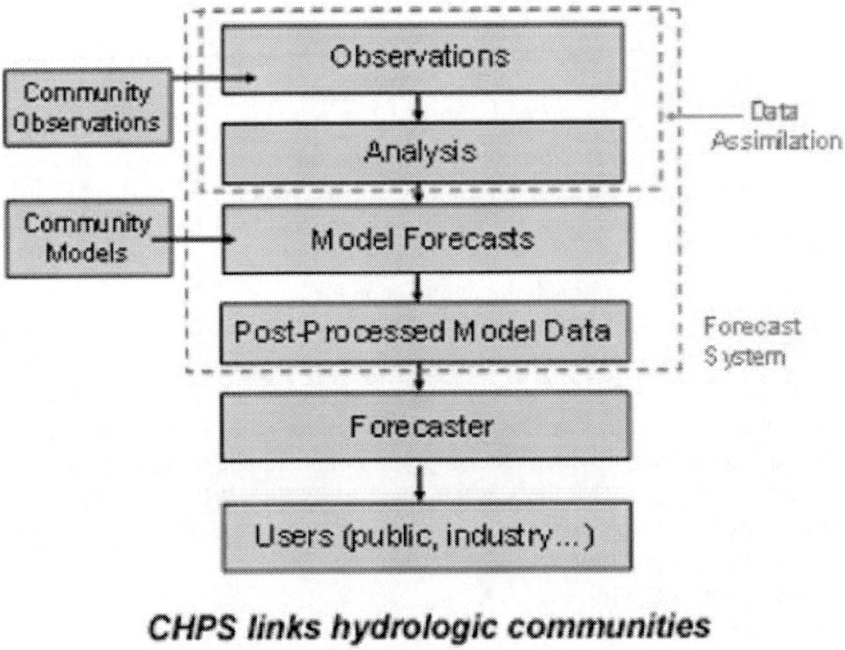

Figure 13. Architecture of the Community Hydrologic Prediction System (National Oceanic and Atmospheric Administration, 2011a).

Hydrologic models that predict watershed response to climate and land-use forcings have been in use for more than 30 years. Although models have improved in their ability to manage and display data, many of the underlying model algorithms remained unchanged from earlier versions. Hydraulic models, model movement of water through aquifers and streams, have improved capacity to predict water and solute movement in complex situations such as coupled river–flood-plain systems, reservoirs, and estuaries. Water-quality models can perform reasonably adequate predictions of temperature, nutrient, dissolved-oxygen, and salinity in surface waters. Groundwater models also can predict salinity, temperature, and many dissolved constituents adequately. Prediction of other contaminants is more challenging, particularly contaminants that sorb and desorb to particulates. As suggested in **Review Element 5**, development and integration of models can be used to design follow-up field studies and also to inform policy decisions. Conversely, models can be used to optimize sampling programs and the development of monitoring networks.

Sustained, coordinated monitoring and ongoing analysis of data are needed to develop and implement effective structural and operational adjustments in response to observed system changes. A key barrier to applying climate change vulnerability assessment tools to ecosystems is the need to ensure that information and models that are used are appropriate for the scale of the decision (water body, watershed, regional, national). Adaptive management requires institutions to reevaluate project and program performance and to revise decisions based on the most recent information, which requires a commitment to ongoing monitoring.

There are several efforts currently underway to assist in meeting these challenges, including the Water Utilities Climate Alliance (WUCA) and the Integrated Water Resources Science and Service (IWRSS). WUCA (http://www.wucaonline. org/html/) is a consortium of 10 large water utilities that was formed to provide leadership and collaboration on climate- change challenges facing the Nation's water managers. The IWRSS currently is a partnership of NOAA, the Corps of Engineers, and USGS, agencies with complementary operational missions in water science, observation, prediction and management (Figure 14).

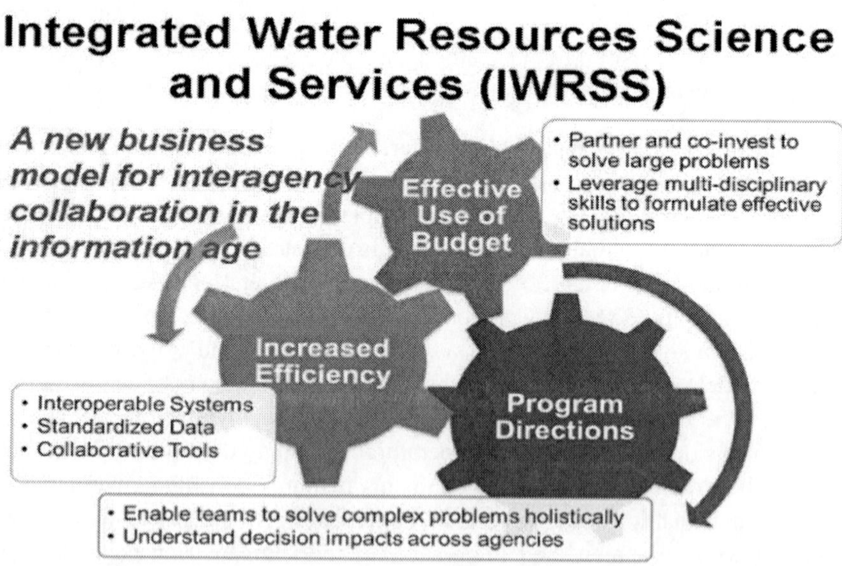

Figure 14. Overview of the Integrated Water Resources Science and Service Business Model.

The National Science and Technology Council's (NSTC) Subcommittee on Water Availability and Quality (SWAQ) is the one venue in which all Federal agencies with any responsibility (or interest) in water issues gather regularly to exchange information. The SWAQ advises and assists the NSTC on policies, procedures, plans, issues, scientific developments, and research needs related to the availability and quality of water resources of the United States within the context of the global hydrologic cycle. The SWAQ focuses on science issues and associated policy options related to research and improvements in technology to advance the goal of ensuring a safe and sustainable supply of water in the United States in order to preserve the capacity to support societal and ecological needs, particularly in the context of a changing climate.

Finding 24: Existing Coordination. Significant ecological and social benefits are expected from enhanced coordination of water-related data collection and integration of activities across the Federal government and with non- Federal partners.

The Advisory Committee on Water Information (ACWI; http://acwi.gov) is a Federal Advisory Committee that represents the interests of water-information users and professionals. ACWI represents the interests of water-information users and professionals in advising the Federal Government on Federal water-information programs and their effectiveness in meeting the Nation's water-information needs. ACWI supports a number of subcommittees, including the Subcommittee on Spatial Water Data and the Sustainable Water Resources Roundtable, both of which are relevant to findings presented in the report. ACWI provides a neutral forum for collaboration on the collection, storage, management, analysis, and dissemination of climate-relevant water information.

Related to the coordinated use and application of water resources data and information for natural resources management, a committee was formed in 2010, in coordination with the Interagency Climate Change Adaptation Task Force and composed of Federal, state, and tribal partners, with input from many others, to develop a collective national adaptation strategy for fish, wildlife, and plant resources and their habitats. The committee's objective is to produce a National Fish, Wildlife, and Plants Climate Adaptation Strategy (NFWPCAS, 2011) by 2012 that identifies and defines principles and meth ods to maintain key terrestrial, freshwater, and marine ecosystems and functions needed to sustain fish, wildlife and plants. This blueprint for common action will outline needed scientific support, policy and legal frameworks; best management practices; processes for integration and communication; and a framework for implementing these approaches at regional and local scales.

Neither the SWAQ, ACWI, nor the NFWPCAS team currently are constituted to address all of the water-related challenges facing the Nation. These two entities, however, represent opportunities for improved coordination across Federal agencies and with Tribal, state, and local groups.

NEXT STEPS

Effective management of the Nation's water resources will require meaningful action to address many of the short-comings that were identified in this report on hydroclimatic observational and modeling systems. Moreover, continued vigilance and effort will be required to sustain and enhance vital observation and modeling systems that currently are in place and operating effectively over a period of years. Federal agencies should consider amending existing programs and policies to incorporate these actions whenever possible and should consider these actions in future budget planning.

Finding 25: Existing Coordination Mechanisms. Federal agency actions towards improving data integration should continue and be accelerated through the Subcommittee on Water Availability (SWAQ) and the Advisory Committee on Water Information (ACWI). SWAQ should be strengthened to better promote cross-agency communication and additional coordination opportunities. The Advisory Committee on Water Information provides a useful, and perhaps under-utilized forum for collaboration on the collection, storage, management, analysis, and dissemination of climate-relevant water information between Federal and Tribal, state, and local agencies.

This report does not prioritize among the needs that were identified, but prioritization will be required as Congress and the water-resources community begin to deliberate on these findings. Certainly, high priority should be given to maintaining existing capabilities, as measurements of hydroclimatic processes that are not made can never be recovered, creating unrecoverable gaps in the systematic record. Findings in this report that are supported by other documents (for example, the National Monitoring Network and the National Groundwater Monitoring Network) and those groups also deserve priority consideration.

Key next steps for Federal agencies to implement the findings of this report are described below.

1. **Strengthen observational data systems for fresh-water resources and climate change.**
    - Strengthen existing efforts, including the Water Census, through enhancements to the hydroclimatic observational network identified in **Review Element 1** of this report. Information that is critical from a health, safety, and welfare perspective should be given priority, while considering the data needs of land, water and environmental resource managers. Effects on water resources from energy extraction and production, and carbon sequestration (both geologic and biological) need increased emphasis.
    - Conduct ongoing and sustained analyses of hydroclimatic data to identify emerging trends and patterns and to develop new insight into hydroclimatic variability.
2. **Prioritize observational systems that fill important gaps in understanding water supply reliability.**
    - Enhance collection of water-use information, including provision of timely information on withdrawals and return flows (quantity and quality) from surface and groundwater resources and information on withdrawals and consumptive use by sector.
    - Implement the proposed National Streamflow Information Program and the National Groundwater Monitoring Network, both of which are authorized in PL 111-11.
    - Develop and implement a national lake/reservoir level and contents data network.
    - In partnership with private industry, conduct research on new monitoring technologies, including sensors, data transmission, automated quality assurance, and remote-sensing technologies.
3. **Improve water-quality and ecosystem monitoring systems.**
    - Implement the National Water-quality Monitoring Network, which is supported by the Advisory Committee on Water Information and is consistent with the National Ocean Council's Strategic Action Plans.

- Enhance interagency efforts and support States to monitor and improve mapping of wetland areas and habitat quality on a seasonal basis.
- Implement a waterborne disease tracking network, including all appropriate ancillary data.

4. **Strengthen links between hydroclimatic observational data systems and climate models; improve data management, acquisition, analysis, and reporting.**
    - Link monitoring, observational systems, climate model outputs and other data systems, to update and improve hydroclimatic statistics that support high-priority water management decisions (particularly related to water supply reliability and quality).
    - Build on the initial foundation established by the Integrated Water Resources Sciences and Services activity to expand and encourage the use of consistent data standards across agencies and with non-governmental partners and other ways to integrate existing data into more comprehensive water information systems.
    - Develop new and improved models (both statistical and deterministic) for assessing hydroclimatic data, developing design conditions, and forecasting likely future conditions for expected scenarios.
    - Develop guidance for water managers on appropriate use of probabilistic projections and model outputs.
5. **Support the establishment of an interagency climate data portal and provide access to high priority water-related datasets.**
    - Promote interagency coordination of diverse data and define the architecture of data systems for freshwater resources to facilitate improved access to these data through a single portal. As a part of the portal, provide for user feedback, including recommendations for improvements or modifications.
6. **Strengthen coordination to improve the quality and accessibility of freshwater data systems including technical outreach and support to stakeholders and decision-makers.**
    - Request that the Subcommittee on Water Availability and Quality (SWAQ) monitor progress on implementing the findings and recommendations of this report and provide annual updates to the National Science and Technology Council and member agencies, and to non-Federal partners.

- Promote interagency coordination and cooperation to implement the National Water Census (http://water.usgs.gov/wsi/) and Integrated Water Resources Science and Services (National Oceanic and Atmospheric Administration, 2009).
- Encourage the Advisory Committee on Water Information, an existing Federal Advisory Committee, to establish a new subcommittee or other appropriate mechanism, to solicit and consider input from the public and stakeholders on matters related to freshwater resources and a changing climate and relay these views to Federal water data program managers and the SWAQ.
- Fully engage States, Tribes, local agencies, and interstate organizations, most of whom have water-resources management responsibilities, in the implementation of the findings in this report.

ACRONYMS

Acronym	Definition
ACWI	Advisory Committee on Water Information
ALI	Hyperion and Advanced Land Imager
ASOS	Automated Surface Observing System
ASTAR	Advanced Spaceborne Thermal Emission and Reflection Radiometer
ATTAINS	Assessment Total Daily Maximum Load Tracking and Implementation System
AVHRR	Advanced Very High Resloution Radiometer
BMPs	Best Management Practices
BoR	Bureau of Reclamation
CCAWWG	Climate Change and Water Working Group
CDC	Centers for Disease Control and Prevention
CEN	Climate Effects Network
CENRS	Committee on Environmental and Natural Resources and Sustainability
CHS	Canadian Hydrographic Service
CHPS	Community Hydrologic Prediction System
COOP	Cooperative Observer Program
CRN	Climate Response Network

(Continued)

Acronym	Definition
CUAHSI	Consortium of Universities for the Advancement of Hydrologic Science Inc.
DOQQ	Digital Orthophoto Quadrangles
ELGs	Effluent Limitation Guidelines
EMAP	Environmental Monitoring and Assessment Program
FGDC	Federal Geographic Data Committee
FRS	Farm and Ranch Survey
GCMs	Global Climate Models
HIS	Hydrologic Information System
IWRSS	Integrated Water Resources Science and Services
ICCATF	Interagency Climate Change Adaptation Task Force
IOOS®	U.S. Integrated Ocean Observing System
IPCC	Intergovernmental Panel on Climate Change
LDCM	Landstat Data Continuity Mission
LIDAR	Light Detection and Ranging
MODIS	Moderate Resolution Imgaing Spectroradiometer
MMS	Modular Modeling System
NAPP	National Aerial Photography Program
NACC	National Asessment on Climate Change
NARS	National Aquatic Research Survey
NASA	National Areonautics and Space Agency
NASQAN	National Stream Quality Accounting Network
NAWQA	National Water-quality Assessment Program
NCDC	National Climate Data Center
NCCR	National Coastal Commission Report
NEP	National Estuary Programs
NEXRAD	Next-Generation Radar
NFDM	Nations' Freshwater Data and Models
NFWPCAS	National Fish, Wildlife, and Plants Climate Adaptation Strategy
NHANES	National Health and Nutrition Examination Survey
NHAP	National High Altitude Photography
NOAA	National Oceanic and Atmospheric Administration
NOHRSC	National Operational Hydrologic Remote Sensing Center
NPDES	National Pollutant Discharge Elimination System
NPS	National Park Service

Acronym	Definition
NRCS	Natural Resources Conservation Service
NSA	National Snow Analysis
NSF	National Science Foundation
NSIP	National Streamflow Information Program
NSTC	National Science and Technology Council
NSQAP	National Stream Quality Assessment Program
NWIS	National Water Information System
NWS	National Weather Service
NWUIP	National Water-Use Information Program
OSTP	U.S. Office of Science and Technology Policy
RESSED	Reservoir Sedimentation Database
RLLN	River and Lake Level Network
RUSHCN	Regional U.S. Historical Climate Network
SCAN	Soil Climate Analysis Network
SDWA	Safe Drinking Water Act
SMAP	Soil Moisture Active Passive
SNOTEL	Snowpack Telemetry
SRTM	Shuttle Radar Topography Mission
SS-WSF	Snow Survey and Water Supply Forecasting
STORET	Storage and Retrival Data Warehouse
SWAQ	Subcommittee on Water Availability and Quality
TMDLs	Total Maximum Daily Loads
TVA	Tennessee Valley Authority
UIC	Underground Injection Control
USACE	U.S. Army Corps of Engineers
CRN	U.S. Climate Reference Network
USDA–ARS	U.S. Department of Agriculture–Agriculture Research Service
USDA–FS	U.S. Department of Agriculture–Forest Service
USEPA	U.S. Environmental Protection Agency
USFWS	U.S. Fish and Wildlife Service
USGS	U.S. Geological Survey
USGCRP	U.S. Global Change Research Program
WSA	Wadeable Streams Assessment
WaterML	Water Mark-up Language
WQS	Water Quality Criteria and Standards
WRCCW	Water Resources and Climate Change Workgroup
WHCEQ	White House Council on Environmental Quality

Appendix A. SECURE Water Act: SEC. 9506. Climate Change and Water IntraGovernmental Panel

123 STAT. 1338 PUBLIC LAW 111–11—MAR. 30, 2009

SEC. 9506. CLIMATE CHANGE AND WATER INTRA-GOVERNMENTAL PANEL.

(a) ESTABLISHMENT.—The Secretary and the Administrator shall establish and lead a climate change and water intragovernmental panel—

(1) to review the current scientific understanding of each impact of global climate change on the quantity and quality of freshwater resources of the United States; and

(2) to develop any strategy that the panel determines to be necessary to improve observational capabilities, expand data acquisition, or take other actions—

(A) to increase the reliability and accuracy of modeling and prediction systems to benefit water managers at the Federal, State, and local levels; and

(B) to increase the understanding of the impacts of climate change on aquatic ecosystems.

(b) MEMBERSHIP.—The panel shall be comprised of—

(1) the Secretary;

(2) the Director;

(3) the Administrator;

(4) the Secretary of Agriculture (acting through the Under Secretary for Natural Resources and Environment);

(5) the Commissioner;

(6) the Secretary of the Army, acting through the Chief of Engineers;

(7) the Administrator of the U.S. Environmental Protection Agency; and

(8) the Secretary of Energy.

(c) REVIEW ELEMENTS.—In conducting the review and developing the strategy under subsection (a), the panel shall consult with State water resource agencies, the Advisory Committee, drinking water utilities, water research organizations, and relevant water user, environmental, and other nongovernmental organizations—

(1) to assess the extent to which the conduct of measures of streamflow, groundwater levels, soil moisture, evapotranspiration rates, evaporation rates, snowpack levels, precipitation amounts, flood risk, and glacier mass is necessary to improve the understanding of the Federal Government and the States with respect to each impact of global climate change on water resources;

(2) to identify data gaps in current water monitoring networks that must be addressed to improve the capability of the Federal Government and the States to measure, analyze, and predict changes to the quality and quantity of water resources, including flood risks, that are directly or indirectly affected by global climate change;

(3) to establish data management and communication protocols and standards to increase the quality and efficiency by which each Federal agency acquires and reports relevant data;

(4) to consider options for the establishment of a data portal to enhance access to water resource data—

(A) relating to each nationally significant freshwater watershed and aquifer located in the United States; and

(B) that is collected by each Federal agency and any other public or private entity for each nationally significant freshwater watershed and aquifer located in the United States;

(5) to facilitate the development of hydrologic and other models to integrate data that reflects groundwater and surface water interactions; and

(6) to apply the hydrologic and other models developed under paragraph (5) to water resource management problems identified by the panel, including the need to maintain or improve ecological resiliency at watershed and aquifer system scales.

(d) REPORT.—Not later than 2 years after the date of enactment of this Act, the Secretary shall submit to the appropriate committees of Congress a report that describes the review conducted, and the strategy developed, by the panel under subsection (a).

(e) DEMONSTRATION, RESEARCH, AND METHODOLOGY DEVELOPMENT PROJECTS.—

(1) AUTHORITY OF SECRETARY.—The Secretary, in consultation with the panel and the Advisory Committee, may provide grants to, or enter into any contract, cooperative agreement, interagency

agreement, or other transaction with, an appropriate entity to carry out any demonstration, research, or methodology development project that the Secretary determines to be necessary to assist in the implementation of the strategy developed by the panel under subsection (a)(2).

(2) REQUIREMENTS.—

(A) MAXIMUM AMOUNT OF FEDERAL SHARE.—The Federal share of the cost of any demonstration, research, or methodology development project that is the subject of any grant, contract, cooperative agreement, interagency agreement, or other transaction entered into between the Secretary and an appropriate entity under paragraph (1) shall not exceed $1,000,000.

(B) REPORT.—An appropriate entity that receives funds from a grant, contract, cooperative agreement, interagency agreement, or other transaction entered into between the Secretary and the appropriate entity under paragraph (1) shall submit to the Secretary a report describing the results of the demonstration, research, or methodology development project conducted by the appropriate entity.

(f) AUTHORIZATION OF APPROPRIATIONS.—

(1) IN GENERAL.—There is authorized to be appropriated to carry out subsections (a) through (d) $2,000,000 for each of fiscal years 2009 through 2011, to remain available until expended.

(2) DEMONSTRATION, RESEARCH, AND METHODOLOGY DEVELOPMENT PROJECTS.—There is authorized to be appropriated to carry out subsection (e) $10,000,000 for the period of fiscal years 2009 through 2013, to remain available until expended.

APPENDIX B. EXAMPLES OF EXISTING NETWORKS THAT PRODUCE OBSERVATIONAL DATA RELEVANT TO CLIMATE CHANGE AND WATER RESOURCES

- Streamgage network: approximately 7,500 streamgages (ref ACWI) where observations are made at specific time intervals that range from

15-minutes to 1-hour with data transmitted via satellite for dissemination to the public via the Internet

- Groundwater monitoring network
- Groundwater Climate Response Network
- Precipitation: Cooperative Observer Program (COOP), Automated Surface Observing System (ASOS), Snow Survey and Water Supply Forecasting (SS-WSF), and Snowpack Telemetry (SNOTEL), U.S. Climate Reference Network (CRN) and the Regional U.S. Historical Climate Network (RUSHCN)
- Water quality: National Water Information System (NWIS) includes surface water, groundwater, and water-quality data spanning all 50 States, plus border and territorial sites, and include data from as early as 1899 to present.
- Land remote sensing: Remote sensing products are developed by NASA, NOAA, USGS, and other Federal, State, and private sector partners using satellites and aircraft monitor the Earth. The satellite imagery and aerial photographs are archived and used to compare current conditions with historical land features. Data can be interpreted over time and space to assess the impact of natural disasters, climate, and other global changes. A summary of the components of the USGS Land and Remote Sensing (LRS) is provided in Appendix C and can be accessed on-line at http://remotesensing. usgs.gov/.
- National Stream Quality Accounting Network (NASQAN): concentrations and loads of selected constituents delivered by major rivers to the coastal waters of the United States and selected inland sub-basins in priority river basins to determine the sources and relative yields of constituents within these basins. These priority basins have significant management interest in reducing delivery of constituents that contribute to adverse conditions in receiving waters. Other objectives include monitoring for climate change and describing long-term trends in the loads and concentrations of select constituents at key locations.
- Nationally notifiable disease surveillance system
- National Health and Nutrition Examination Survey (NHANES): Collection of data on the health and nutritional status of a nationally representative sample of about 5,000 persons each year.
- Benchmark glacier monitoring http://ak.water.usgs.gov/glaciology/

Appendix C. Examples of Current Databases That Incorporate Information Relevant to Climate Change and Water Resources

Observational Data

- The Natural Resources Inventory (NRI): database of the status, condition, and trends of land, soil, water, and related resources on the Nation's non-Federal lands, which is maintained by USDA National Resources Conservation Service (NRCS). The data are collected yearly and continually updated and used by natural resource managers; policymakers and analysts; consultants; the media; other Federal agencies; State governments; universities; environmental, commodity, and farm groups; and the public. Much of the farm-scale data are not publicly available.
- National Water Information System (NWIS) database: database containing streamflow, water-quality, ground-water levels and quality, algal composition, quality of bed material, and composition of animal tissue.
- National Ambient Water-quality Assessment Program (NAWQA): Database of physical, chemical, and biological data from specific sites across the U.S.
- Storage and Retrieval Data Warehouse (STORET): repository for water quality, biological, and physical data and is used by State environmental agencies, EPA and other Federal agencies, universities, private citizens, and many others.
- Assessment Total Daily Maximum Load Tracking and Implementation System (ATTAINS): provides access to information on impaired waters.
- NOAA's National Weather Service interactive website (http://water.weather.gov/ahps/): provides easy access to data on weather, rainfall, snow, flood-stage status, hydrographs, digital tools to generate flood inundation maps
- National Streamflow Information database (NSIP)
- National Climatic Data Center: Maintains a broad range of national and global climate data sets, as well as model archives.

Aerial Imagery Maintained by the Land Remote Sensing Program

- National Aerial Photography Program (NAPP) - The National Aerial Photography Program is an interagency Federal effort coordinated by the USGS, which uses NAPP products to revise maps.
- National High Altitude Photography (NHAP) - The National High Altitude Photography (NHAP) program, which was operated from 1980-1989, was coordinated by the U.S. Geological Survey as an interagency project to eliminate duplicate photography in various Government programs.
- Digital Orthophoto Quadrangles (DOQQ) - Digital images of aerial photos which combine the image characteristics of the photo with the georeferenced qualities of a map
- Light Detection and Ranging (LIDAR) - LIDAR is an optical remote sensing technology that measures properties of scattered light to find range and/or other information of a distant target.
- Aircraft scanners - Digital imagery acquired from several multispectral scanners on board NASA ER–2, NASA C–130B, and NASA Learjet aircrafts
- Historical, commercial, and Scientific Committee on Antarctic Research (SCOR) imagery

Satellite Imagery Maintained by Land Remote Sensing Program

- Advanced Spaceborn Thermal Emission and Reflection Radiometer (ASTAR) - High-resolution multispectral data from the *Terra* satellite
- Advanced Very High Resolution Radiometer (AVHRR) - 1-km multispectral data from the NOAA satellite series
- Hyperion and Advanced Land Imager (ALI) - 10- to 30-meter multispectral and hyperspectral data from the Earth Observing-1 (EO-1) Extended Mission
- LANDSAT - Landsat satellites have been providing multispectral images of the Earth continuously since the early 1970's.
- Landsat ETM+ (Enhanced Thematic Mapper Plus)–High-resolution (15- to 60-meter) multispectral data from Landsat 7 (1999 to present)

- Landsat MSS (Multispectral Scanner) - 80-meter multispectral data from Landsats 1 to 5 (1972 to 1992).
- Landsat TM (Thematic Mapper) - 30- to 120-meter multispectral data from Landsat 4 and 5 (1982 to present).
- Landsat Data Continuity Mission (LDCM) - Multispectral data from the proposed Landsat Data Continuity Mission.
- Moderate Resolution Imaging Spectroradiometer (MODIS) - Moderate-resolution (250- to 1000-meter) multispectral data from the Terra Satellite (2000 to present) and Aqua Satellite (2002 to present).
- Shuttle Radar Topography Mission (SRTM) - SRTM data are used to generate a digital topographic map of the Earth's land surface with data points spaced every 1 arc second of latitude and longituded for the United States (approximately every 30 meters).
- SIR-C (Spaceborne Imaging Radar C-band) - Imaging radar data (C-band and L-band) from two Space Shuttle missions (1994).
- *Declassified Satellite Images - Category 1* - Photographic imagery from the CORONA, ARGON and LANYARD satellites (1959 to 1972).
- *Declassified Satellite Images - Category 2* - Photographic imagery from KH-7 Surveillance and KH-9 Mapping system (1963 to 1980).
- *USGS Commercial Data Purchases (UCDP) Imagery* - The UCDP Imagery Collection consists of imagery from several commercial vendors. The UCDP supports the *Commercial Remote Sensing Space Policy* (CRSSP) by providing data to qualified users, primarily U.S. Federal agencies, at no cost for File Transfer Protocol (FTP) downloads or at a nominal cost for media.

APPENDIX D. MAJOR CONTRIBUTORS TO THIS REPORT

Jerad Bales, U.S. Geological Survey
David Bylsma, U.S. Environmental Protection Agency
Judsen Bruzgul, Council on Environmental Quality
John Ingram, National Weather Service
Kathy Jacobs, U.S. Global Change Research Program
Rodney Knight, U.S. Geological Survey
Audrey Levine, U.S. Environmental Protection Agency

Trey Lewis, Council on Environmental Quality
Jeff Peterson, Council on Environmental Quality

REFERENCES

Ackerman, K. V., Mixon, D. M., Sundquist, E. T., Stallard, R. F., Schwarz, G. E. & Stewart, D. W. (2009). RESIS—An updated version of the original Reservoir Sedimentation Survey Information System (RESIS) database, *U.S. Geological Survey Data Service.*, 434 p., available at (http://pubs.usgs.gov/ds/ds434).

Advisory Committee on Water Information. (2010). available at http://acwi.gov/.

Alexander, R. B., Slack, J. R., Ludtke, A. S., Fitzgerald, K. K. & Schertz, T. L. (1997). Data from selected U.S. Geological Survey National Stream Water-Quality Monitoring Networks (WQN) on CD-ROM: *U.S. Geological Survey Fact Sheet*, FS-013-97, available at http://pubs.usgs.gov/fs/FS-013-97/.

Allen, R. G., Tasumi, M. & Trezza, R. (2007). Satellite-based energy balance for mapping evapotranspiration with internalized calibration (METRIC)—Model: *Journal of Irrigation and Drainage Engineering*, v. *133*, no. 4, p. 380–394.

AmeriFlux. (2011). Site and Data Exploration System, available at http://ameriflux.ornl.gov/.

Anderson, M. C., Kustas, W. P. & Norman, J. M. (2007). Upscaling flux observations from local to continental scales using thermal remote sensing: *Agronomy Journal*, v. *99*, p. 240–254, doi:10.2134/agronj2005.0096S.

Arthington, A. H., Bunn, S. E., Poff, N. L. & Naiman, R. J. (2006). The challenge of providing environmental flow rules to sustain river ecosystems: *Ecological Applications*, v. *16*, no. 4, p. 1311–1318.

Bae, D.-H., Jung, I.-W. & Lettenmaier, D. P. (2011). Hydrologic uncertainties in climate change from IPCC AR4 GCM simulations of th Chungju Basin, Korea: *Journal of Hydrology*, v. *401*, p. 90–105.

Beller-Simms, N., Ingram, H., Feldman, D., Mantua, N., Jacobs, K. L. & Waple, A. (2008). Decision-Support Experiments and Evaluations Using Seasonal to Interannual Forecasts and Observational Data: A Focus on Water resources, Synthesis and Assessment, Product 5.3.

Brekke, L. D., Kiang, J. E., Olsen, J. R., Pulwarty, R. S., Raff, D. A., Turnipseed, D. P., Webb, R. S. & White, K. D. (2009). Climate Change and Water Resources Management—A Federal Perspective: *U.S. Geological Survey Circular*, *1331*, 65 p., available at http://pubs.usgs.gov/circ/1331/

Brekke, L. D. (2011). Addressing Climate Change in Long-Term Water Resources Planning and Management: User Needs for Improving Tools and Information, U.S. Army Corps of Engineers CWTS-10-02, available at http://www.usbr.gov/ climate/userneeds/docs/LTdoc.pdf

Carlisle, D. M., Wolock, D. M. & Meador, M. R., (2010). Alteration of streamflow magnitudes and potential ecologic consequences—A multiregional assessment, Front Ecol. Environ, 2010, available at http://water.usgs.gov/nawqa/ pubs/Carlisleetal_FLowAlterationUS.pdf.

Centers for Disease Control and Prevention. (2011). *Climate Change and Public Health*, available at http://www.cdc.gov/ climatechange/.

CHRS-PERSIANN-CCS. (2011). Precipitation, available at http://fire.eng.uci.edu/CCS/.

Congressional Research Service. (2009). 35 years of Water Policy: *The 1973 National Water Commission and Present Challenges*: Report R40573, 68 p.

Consortium of Universities for the Advancement of Hydrologic Science, Inc. (2011). Hydrologic Information System, available at http://www.cuahsi.org/his.html.

Committee on Environment Natural Resources, and Sustainability. (2010). available at http://www.whitehouse.gov/administration/eop/ostp/nstc/committees/cenrs.

The Council on Environmental Quality. (2003). Modernizing NEPA Implementation: *NEPA Task Force report*, Washington, D.C., 122 p., available at http://ceq.hss.doe.gov/ntf/ report/finalreport.pdf.

The Council on Environmental Quality (CEQ). (2010). *Interagency Climate Change Adaptation Task Force*, *Progress Report*, accessed October 5, 2010, at http:// www.whitehouse.gov/administration/eop/ceq/initiatives/adaptation.

Cunningham, W. L., Geiger, L. H. & Karavitis, G. A. (2007). U.S. Geological Survey Ground-Water Climate Response Network: *U.S. Geological Survey Fact Sheet*, 2007–3003, available at http://pubs.usgs.gov/fs/2007/3003/.

Dooley, J. J., Dahowski, R. T., Davidson, C. L., Wise, M. A., Gupta, N., Kim, S. H. & Malone, E. L. (2006). Carbon Dioxide Capture and Geologic

Storage—A Core Element of a Global Energy Technology Strategy to Address Climate Change: *A Technology Report from the Second Phase of the Global Energy Technology Strategy Program*, PNNL-15296, Battelle, Joint Global Change Research Institute, Pacific Northwest National Laboratory, College Park, Md., available at http://www.pnl.gov/gtsp/docs/ccs_report.pdf.

Electric Power Research Institute (EPRI). (2002). Water and Sustainability (Volume 3): U.S. Water Consumption for Power Production—*The Next Half Century*, Report Number 1006786, Electric Power Research Institute, Palo Alto, Calif., available at http://www.epri.com (This report is downloadable from the EPRI website.)

Furniss, M. J., Staab, B. P., Hazelhurst, S., Clifton, C. F., Roby, K. B., Ilhadrt, B. L., Larry, E. B., Todd, A. H., Reid, L. M., Hines, S. J., Bennet, K. A., Luce, C. H. & Edwards, P. J. (2010). Water, climate change, and forests—Watershed stewardship for a changing climate: *General Technical Report* PNW-GTR-812, Portland, Oreg.

Hapke, C. J., Himmelstoss, E. A., Kratzmann, M., List, J. H. & Thieler, E. R. (2010). National assessment of shoreline change; historical shoreline change along the New England and Mid-Atlantic coasts: *U.S. Geological Survey Open-File Report*, 2010–1118, 57 p., available at http://pubs.usgs.gov/ of/2010/1118/.

Hatton, T. J., Moore, S. J. & Reece, P. H. (1995). Estimating stand transpiration in a Eucalyptus populnea woodland with the heat pulse method: measurement errors and sampling strategies: *Tree Physiology*, v. *15*, no. 4, p. 219–227, doi: 10.1093/treephys/15.4.219.

H. John Heinz III Center. (2002). The State of the Nation's Ecosystems—Measuring the Lands, Waters, and Living Resources of the United States: The H. John Heinz III Center for Science, Economics, and the Environment, Washington, D.C., Cambridge University Press, 34 p., available at http://www.heinzctr.org/ecosystems/index.htm.

Hutson, S. S., Barber, N. L., Kenny, J. F., Linsey, K. S., Lumia, D. S. & Maupin, M. (2004). Estimated use of water in the United States: *U.S. Geological Survey Circular*, *1268*, 46 p., accessed December 8, 2010, at http://pubs.usgs.gov/circ/2004/circ1268/.

Interagency Working Group on Climate Change and Health. (2010a). A Human health perspective on climate change: *Research Triangle Park, N.C.*, The National Institute of Environmental Health Sciences, 70 p., available at http:// www.cdc.gov/climatechange/pubs/HHCC_Final_508.pdf.

Interagency Climate Change Adaptation Task Force. (2010b). Progress Report of the Interagency Climate Change Adaptation Task Force: Recommended Actions in Support of a National Climate Change Adaptation Strategy: The White House Council on Environmental Quality, http://www.whitehouse.gov/sites/default/files/microsites/ceq/Interagency-Climate-Change-Adaptation-Progress-Report. pdf.

Intergovernmental Panel on Climate Change (IPCC). (2001). Climate Change 2001: Contribution of Working Group II to the Third Assessment Report of the Intergovernmental Panel on Climate Change, Impacts, Adaptation, and Vulnerability. [McCarthy, J.J., Canziani, O.F., Leary, N.A., Dokken, D.J., and White, K.S., eds.], Cambridge University Press, Cambridge, United Kingdom and New York, N.Y., 1,032 p., available at http://www.grida.no/publications/other/ipcc_ tar/.

Intergovernmental Panel on Climate Change (IPCC). (2005). IPCC Special Report on Carbon Dioxide Capture and Storage, prepared by Working Group III of the Intergovernmental Panel on Climate Change [Metz, B.O., Davidson, H., de Coninck, M., Loos, L., Meyer, L.A., eds.]. Cambridge University Press, Cambridge, United Kingdom and New York, N.Y., USA, available at http://www.ipcc.ch/publications_and_data/publications_and_data_reports. shtml.

Intergovernmental Panel on Climate Change (IPCC). (2007a). Climate Change 2007: The Physical Science Basis, Contribution of Working Group I to the Fourth Assessment Report of the Intergovernmental Panel on Climate Change [Solomon, S.D., Qin, D., Manning, M., Chen, Z., Marquis, M., Averyt, K.B., Tignor, M.M.B., and Miller, H.L., eds.], Cambridge University Press, Cambridge, United Kingdom and New York, N.Y., 996 p., available at http://ipcc.ch/publications_and_data/ar4/wg1/en/contents.html.

Intergovernmental Panel on Climate Change (IPCC). (2007b). Climate Change 2007: Impacts, Adaptation and Vulnerability, Contribution of Working Group II to the Fourth Assessment Report of the Intergovernmental Panel on Climate Change. [Parry, M.L., Canziani, O.F., Palutikof, J.P., van der Linden, P.J., and Hanson, C.E., eds.], Cambridge University Press, Cambridge, United Kingdom and New York, N.Y., 976 p., available at http://ipcc.ch/publications_and_data/ar4/wg2/en/contents.html.

Intergovernmental Panel on Climate Change (IPCC). (2007c). Climate Change 2007: Mitigation. Contribution of Working Group III to the Fourth Assessment Report of the Intergovernmental Panel on Climate Change. [Metz, B., Davidson, O.R., Bosch, P.R., Dave, R., Meyer, L.A., eds.],

Cambridge University Press, Cambridge, United Kingdom and New York, N.Y., 851 p., available at http://ipcc.ch/publications_ and_data/ar4/wg3/ en/contents.html.

Intergovernmental Panel on Climate Change (IPCC). (2008). *Technical Paper VI*, IPCC Secretariat, Working Group II, Technical Support Unit, Geneva, available at http://www. ipcc.ch/pdf/technical-papers/climate-change-water-en.pdf

Johnson Foundation. (2010). Charting new waters—*A call to action to address U.S. freshwater challenges*: Meridian Institute, Racine, Wisc., 45 p., available at http://www. johnsonfdn.org/sites/default/files/charting-new-waters/ charting-new-waters.pdf.

Karl, T. R., Melillo, J. M. & Peterson, T. C., eds. (2009). Global Climate Change Impacts in the United States, Cambridge University Press, available at http://www. globalchange.gov/what-we-do/assessment/nca-reports/global-climate-change-impacts-in-the-us-2009.

Keeling, R. F. (2008). Recording Earth's Vital Signs: *Science*, vol. *319*, p. 1771–1772.

Kenny, J. F., Barber, N. L., Hutson, S. S., Linsey, K. S., Lovelace, J. K. & Maupin, M. A. (2009). Estimated use of water in the United States in 2005: *U.S. Geological Survey Circular*, *1344*, 52 p.

Lanthrop, R. G. & Love, A. (2007). Vulnerability of New Jersey's coastal habitats to sea level rise: Grant F. *Walton Center for Remote Sensing & Spatial Analysis*, Rutgers University, New Brunswick, N.J., 17 p., available at http://www.crssa.rutgers.edu/projects/coastal/sealevel/report/ Vulnerability_of_New_Jersey_coastal_habitats_v4.pdf.

Larsen, M. C. (2010). Global change and water resources in the next 100 years: American Water Resources 2010 Summer Specialty Conference, San Juan, Puerto Rico.

Lee, X., Massman, W. & Law, B. E. (2004). Handbook of micrometeorology—*A guide for surface flux measurement and analysis*, Dordrecht, Kluwer Academic Press, 250 p.

Markstrom, S. L., Niswonger, R. G., Regan, R. S., Prudic, D. E. & Barlow, P. M. (2008). GSFLOW-Coupled Ground-water and Surface-water FLOW model based on the integration of the Precipitation-Runoff Modeling System (PRMS) and the Modular Ground-Water Flow Model (MODFLOW-2005): *U.S. Geological Survey Techniques and Methods*, *6–D1*, 240 p., available at http://pubs.usgs.gov/tm/tm6d1.

McCabe, G. J. & Wolock, D. M. (2002). A step increase in streamflow in the conterminous United States, *Geophysical Research Letters*, *29*(24), 2185, doi:10.1029/2002GL015999, available at http://www.agu.org/pubs/crossref/2002/2002GL015999.shtml.

Means, E., Laugier, M., Daw, J., Kaatz, L. & Wagge, M. (2010). Decision Support Planning Methods: Incorporating Climate Change Uncertainties into Water Planning, Prepared for Water Utility Climate Alliance, available at http://www.wucaonline.org/assets/pdf/actions_whitepaper_012110.pdf.

Meehan, G. T., III. (2010). Water Data and Monitoring as Indispensable Tools to Manage Water Quality: *Daily Environment Report*, *148*, DEN B-1, Bureau of National Affairs, Inc.

Mielke, E., Anandon, L. D. & Narayanamurti, V. (2010). Water Consumption of Energy Resource Extraction, Processing, and Conversion: Energy Technology Innovation Policy Research Group, Harvard Kennedy School Belfer Center for Science and International Affairs, Discussion Paper #2010-15, *Energy Technology Innovation Policy Discussion Paper Series*, available at http://belfercenter.ksg. harvard.edu/files/ETIP-DP-2010-15-final-4.pdf.

Milly, P. C. C, Betancourt, J., Falkenmark, M., Hirsch, R. M., Kundzewics, Z. W., Lettenmaier, D. P. & Stouffer, R. J. (2008). "Stationarity is dead: Wither water management?" *Science*, v. *319*, no. 5863, 573 p.

Mueller, B., Seneviratne, S. I., Jimenez, C., Corti, T., Hirschi, M., Balsamo, G., Ciais, P., Dirmeyer, P., Fisher, J. B., Guo, Z., Jung, M., Maignan, F., McCabe, M. F., Reichle, R., Reichstein, M., Rodell, M., Sheffield, J., Tueling, A. J., Wang, K., Wood, E. F. & Zhang, Y. (2011). Evaluation of global observations-based evapotranspiration datasets and IPCC AR4 simulations: *Geophysical Research Letters*, v. *38*, doi:10.1029/2010GL046230, 7 p.

Naiman, R. J., Bunn, S. E., Nilsson, C., Petts, G. E., Pinay, G. & Thompson, L. C. (2002). Legitimizing fluvial ecosystems as users of water—An overview: *Environmental Management*, v. *30*, p. 455–467.

National Aeronautic and Space Administration. (2011a). Goddard Space Flight Center, Tropical Rainfall Measuring System (TRMM), available at http://trmm.gsfc.nasa.gov/.

National Aeronautic and Space Administration. (2011b). SMAP—Soil Moisture Active Passive—Mapping soil moisture and freeze/thaw state from Space, available at http:// smap.jpl.nasa.gov/.

National Fish, Wildlife, and Plants Climate Adaptation Strategy. (2011). available at http://www. wildlifeadaptationstrategy.gov/.

National Ocean Council. (2011). *Strategic Action Plans*, available at http://www.whitehouse.gov/administration/eop/ oceans/sap.

National Oceanic and Atmospheric Administration. (2008). Earth System Research Laboratory, Federal Climate Change and Water Working Group (CCAWWG), available at http:// www.esrl.noaa.gov/psd/ccawwg/.

National Oceanic and Atmospheric Administration. (2009). Integrated Water Resources Science and Services, (IWRSS) An Integrated and Adaptive Roadmap for Operational Implementation: Cline, D., compiler, Report Reference #IWRSS-2009-03-02, available at http://www.nohrsc.noaa.gov/~cline/IWRSS/IWRSS_ROADMAP_v1.0.pdf.

National Oceanic and Atmospheric Administration. (2011a). *Satellite and Information Service, U.S. Climate Reference Network (CRN)*, available at http://www.ncdc.noaa.gov/crn/.

National Oceanic and Atmospheric Administration. (2011b). Hydrometeorological Design Studies Center (HDSC), available at http://www.nws.noaa.gov/oh/hdsc/index.html.

National Oceanic and Atmospheric Administration. (2011c). *Current NWS Probable Maximum Precipitation (PMP) Documents and Related Studies*, available at http://www. nws.noaa.gov/oh/hdsc/studies/pmp.html.

National Oceanic and Atmospheric Administration. (2011d). United States Historical Climatology Network, available at http://www.ncdc.noaa.gov/oa/climate/research/ushcn/ushcn. html.

National Oceanic and Atmospheric Administration. (2011e). Cooperative Observer Program (COOP), available at http:// www.nws.noaa.gov/om/coop/ coopmod.htm.

National Oceanic and Atmospheric Administration. (2011f). National Weather Service Radar Operations Center (NEXRAD), available at http://www.roc.noaa.gov/ WSR88D/.

National Oceanic and Atmospheric Administration. (2011g). Satellite and Information Service, *Satellite Applications and Research (STAR)*, *Satellite Rainfall Estimates*, available at http://www.star.nesdis.noaa.gov/smcd/emb/ff/.

National Oceanic and Atmospheric Administration. (2011h). National Weather Service Climate Prediction Center, CPC Morphing Technique ("CMORPH"), available at http:// www.cpc.ncep.noaa.gov/products/janowiak/cmorph_ description.html.

National Oceanic and Atmospheric Administration. (2011i). National Operational Hydrologic Remote Sensing Center (NOHRSC), available at http://www.nohrsc.nws.gov/ technology/.

National Oceanic and Atmospheric Administration. (2011j). Pacific Northwest salmon habitat restoration project tracking database, available at http://www.nwfsc.noaa.gov/research/divisions/cbd/mathbio/projects/pnwrestoration. cfm.

National Oceanic and Atmospheric Administration. (2011k). Community Hydrologic Prediction System (CHPS), available at http://www.nws.noaa.gov/oh/hrl/chps/index.html.

National Research Council (NRC). (2002). Estimating water use in the United States—*A new paradigm for the National Water-Use Information Program*: National Academy Press, Washington, D.C., 176 p.

Natural Resources Conservation Service. (2011). *National Snow Data Management System Initiative*, available at http:// www.wcc.nrcs.usda.gov/publications/Briefing-Book/bb27. html.

National Water-Quality Monitoring Council (NWQMC). (2008). A national water-quality monitoring network for U.S. coastal waters and their tributaries, available at http://acwi. gov/monitoring/network/design/.

Nelson, D., Elmer, H., Held, R., Forsythe, D. & Casey, S. (2011). Laurentian Great Lakes Basin climate change adaptation: Great Lakes Environmental Research Laboratory, *NOAA Technical Memorandum GLERL, 153*, 43 p., available at http://www.doa.state.wi.us/docview.asp?docid= 8525&locid=9.

Nichols, G., Lane, C., Asgari N., Verlander, N. Q. & Charlett, A. (2009). Rainfall and outbreaks of drinking water related disease and in England and Wales: *Journal of Water and Health*, IWA Publishing, v. *07*, no. 1, p. 1–8.

Norris, J. M. (2009). National Streamflow Information Program, implementation status report: *U.S. Geological Survey Fact Sheet*, 2009–3020, 6 p., available at http://pubs.usgs.gov/fs/2009/3020/.

Noyes, P. D., McElweea, M. K., Millera, H. D., Clark, B. W., Van Tiema, L. A., Walcotta, K. C., Erwina, K. N. & Levina, E. D. (2009). The Toxicology of Climate Change: Environmental Contaminants in a Warming World: *Environment International, 35*, 6, p. 971–986.

Poff, N. L. & Zimmerman, J. K. H. (2010). Ecological response to altered flow regimes—A literature review to the inform science and management of environmental flows: *Freshwater Biology*, v. 55, p. 194–205.

Postel, S. & Richter, B. (2003). Rivers for Life: *Managing Water for People and Nature*, Island Press, Washington, D.C.

Remote Automated Weather Stations. (2011). available at http://raws.fam.nwcg.gov/.

Roderick, M. L. & Farquhar, G. D. (2002). The cause of decreased pan evaporation over the past 50 years: *Science*, *298*(5597), 1410-1411.

Scanlon, B. R., Jolly, I., Sophocleous, M. & Zhang, L. (2007). Global impacts of conversions from natural to agricultural ecosystems on water resources—Quantity versus quality: *Water Resources Research*, v. *43*, no. W03437, doi:10.1029/2006WR005486, 18 p.

Stallard, R. F. (1998). Terrestrial sedimentation and the carbon cycle—Coupling weathering and erosion to carbon burial: *Global Biogeochemical Cycles*, v. *12*, no. 2, p. 231–252.

Stephens, G. L., L'Ecuyer, T., Forbes, R. A., Gettlemen, A., Golaz, J.-C., Bodas-Salcedo, A., Suzuki, K., Gabriel, P. & Haynes, J. (2010). Dreary state of precipitation in global models, *Journal of Geophysical Research*, v. 115, no. D24211, doi:10.1029/2010JD014532.

Thieler, E. R., Williams, Jeff & Hammar-Klose, Erika. (1999–2000). *National Assessment of Coastal Vulnerability to Sea-Level Rise*, available at http://woodshole.er.usgs.gov/project-pages/cvi/.

Subcommittee on Groundwater. (2009). A National Framework for Ground-Water Monitoring in the United States: *Advisory Committee on Water Information*, 74 p., available at http://acwi.gov/sogw/pubs/tr/sogw_tr1_ Framework_june_2009_Final.pdf.

Tihansky, A. B. (1999). Sinkholes, west-central Florida, in Galloway, Devin, Jones, D.R., Ingebritsen, S.E., eds., Land subsidence in the United States: *U.S. Geological Survey Circular*, *1182*, p. 121–140., available at http://pubs.er.usgs. gov/publication/cir1182.

U.S. Commission on Ocean Policy. (2004). *An Ocean Blueprint for the 21st Century*: Washington D.C., available at http:// www.oceancommission. gov/.

U.S. Department of Agriculture. (2008). *Farm and Ranch Irrigation Survey*, available at http://www.agcensus.usda.gov/Publications/2007/Online_ Highlights/Farm_and_Ranch_ Irrigation_Survey/index.asp.

U.S. Department of Agriculture. (2011a). Natural Resources Conservation Service (NRCS), *Soil Climate Analysis Network*, available at http://www.wcc.nrcs.usda.gov/scan/.

U.S. Department of Agriculture. (2011b). Natural Resources Conservation Service, Water Supply Outlook for the Western United States, available at http://www.wcc.nrcs.usda. gov/wsf/westwide.html.

U.S. Department of Agriculture. (2011c). Natural Resources Conservation Service (NRCS), *National Snow Analyses*, available at http://www. nohrsc.nws.gov/nsa/.

U.S. Department of Agriculture. (2011d). *The Census of Agriculture*, available at http://www.agcensus.usda.gov/.

U.S. Department of Agriculture. (2011e). Natural Resources Conservation Service (NRCS), National Water and Climate Center, SNOTEL Data & Products, available at http://www. wcc.nrcs.usda.gov/snow/.

U.S. Department of the Interior. (2011). *Water smart clearing-house*, available at http://www.doi.gov/watersmart/html/ index.php.

U.S. Energy Information Administration. (2011). available at http://www.eia.doe.gov/.

U.S. Environmental Protection Agency. (2006). *Wadeable Streams Assessment—A Collaborative Survey of the Nation's Streams*, EPA 841–B–06–002, available at http:// water.epa.gov/type/rsl/monitoring/streamsurvey/index.cfm

U.S. Environmental Protection Agency. (2008). *National Coastal Condition Report III*, EPA 842-R-08-002, available at http://water.gov/type/oceb/assessmonitor/upload/nccr3_entire.pdf

U.S. Environmental Protection Agency. (2009). *An Introduction to Water-quality Monitoring*, available at http://water.epa.gov/type/watersheds/monitoring/monintr.cfm.

U.S. Environmental Protection Agency. (2010a). *National Lakes Assessment—A Collaborative Survey of the Nation's Lakes: Washington, DC, Office of Wetlands, Oceans & Watersheds*, Report EPA 841–R–09–001, http://water.epa. gov/type/lakes/lakessurvey_index.cfm.

U.S. Environmental Protection Agency. (2010b). *Climate Change Indicators in the United States*, EPA 430–R–10–007, 80 p., available at http://www.epa.gov/climatechange/indicators.html.

U.S. Environmental Protection Agency. (2011a). *National Summary of State Information, available* at http://iaspub.epa. gov/waters10/attains_nation_cy.control.

U.S. Environmental Protection Agency. (2011b). *National Summary Water Quality Attainment in Assessed Bays and Estuaries*, available at http://iaspub.epa.gov/waters10/attains_nation_cy.control#total_assessed_waters.

U.S. Environmental Protection Agency. (2011c). *Water-quality Assessment and Total Maximum Daily Loads Information*, available at http://www.epa.gov/waters/ir/.

U.S. Environmental Protection Agency. (2011d). *Storage and Retrieval (STORET) Data Warehouse, Water Quality Exchange (WQX)*, available at http://www.epa.gov/storet/.

U.S. Geological Survey. (2000). Sea level and climate: *U.S. Geological Survey Fact Sheet* 002–00, available at http://pubs.usgs.gov/fs/fs2-00/.

U.S. Geological Survey. (2003). *Principal Aquifers of the 48 Conterminous United States, Hawaii, Puerto Rico, and the U.S. Virgin Islands*, available at http://www-atlas.usgs.gov/ metadata/aquifrp025.html.

U.S. Geological Survey. (2007). *National Water Census*, available at http://water.usgs.gov/wsi/.

U.S. Geological Survey. (2009a). *National Water-Use Information Program*, available at http://water.usgs.gov/watuse/ wunwup.html.

U.S. Geological Survey. (2009b). *Water Information Coordination Program, Advisory Committee on Water Information*, available at http://ida.water. usgs.gov/ressed/.

U.S. Geological Survey. (2011a). *National Water-Quality Assessment (NAWQA) Program*, Sites for Monitoring of Aquatic Ecological Conditions, available at http://water. usgs.gov/nawqa/studies/mrb/mrb_ sites2.html.

U.S. Geological Survey. (2011b). *National Water Information System (NWIS)*, available at http://waterdata.usgs.gov/nwis.

U.S. Geological Survey. (2011c). *River Forecast Centers*, available at http://water.weather.gov/ahps/rfc/rfc.php.

U.S. Geological Survey. (2011d). *National Streamflow Information Program (NSIP)*, available at http://water.usgs.gov/ nsip/.

U.S. Geological Survey. (2011e). *National Stream Quality Accounting Network Objectives and Subnetworks*, available at http://water.usgs.gov/ nasqan/objectives_subnets.html.

U.S. Geological Survey. (2011f). Groundwater Watch, Climate Response *Network (CRN)*, available at http:// groundwaterwatch.usgs.gov/Net/ OGWNetwork. asp?ncd=crn.

U.S. Geological Survey. (2011g). National Stream Quality Accounting *Network (NASQAN)*, available at http://water. usgs.gov/nasqan/.

U.S. Geological Survey. (2011h). *Hydrologic unit maps*, available at http://water.usgs.gov/GIS/huc.html.

U.S. Geological Survey. (2011i). *National Streamflow Information Program Goals*, available at http://water.usgs.gov/nsip/ goals9.html.

U.S. Geological Survey. (2011j). *National Atlas*, available at http://nationalatlas.gov/.

U.S. Global Change Research Program. (2011). *National Climate Assessment*, available at http://www.globalchange.gov/what-we-do/ assessment.

U.S. Government Accountability Office. (2004). *Watershed management—Better coordination of data collection efforts needed to support key decisions*, Washington, D.C., Report GAO–04–382, 155 p.

U.S. Government Accountability Office. (2009). *Energy-water nexus: improvements to Federal used data would increase understanding of trends in power plant use*: Washington, D.C., Report GAO–10–23, 72 p.

Villarini, G., Smith, J. A., Serinaldi, F. & Ntelekos, A. A. (2011). Analyses of seasonal and annual maximum daily discharge records for central Europe: *J. of Hydrology*, 399:299-312. Vautard, R., Cattiaux, J., Yiou, P., Thepaut, J.-N. & Ciais, P. (2010). Northern Hemisphere atmospheric stilling partly attributed to an increase in surface roughness: *Nature Geoscience*, v. *3*, p. 756–761, doi:10.1038/ngeo979.

Vinukollu, R. K., Wood, E. F., Ferguson, C. R. & Fisher, J. B. (2011). Global estimates of evapotrapnsiration for climate studies using multi-sensor remote sensing data: Evaluation of three process-based approaches: *Remote Sensing of Environment*, v. *115*, p. 801–823.

Wada, Y., van Beek, L., van Kempen, C., Reckman, J., Vasak, S. & Bierkens, M. (2010). Global depletion of groundwater resources: *Geophysical Research Letters*, v. *37*, L20402; doi:10.1029/2010GL044571.

Water Environment Federation. (2010). *Protecting Water Resources and Infrastructure from the Impacts of Climate Change: WEF Position Paper*, 8 p., available at http:// www.wef.org/GovernmentAffairs/PolicyPosition Statement/ ProtectingWRClimateChange.

Williams, K., Pinzon, Z. S., Stumpf, R. P. & Raabe, E. A. (1999). Sea-level rise and coastal forests on the Gulf of Mexico: *U.S. Geological Survey Open-File Report*, 99–441, 87 p., available at http://coastal.er.usgs.gov/wetlands/ofr99-441/OFR99-441.pdf.

Winter, T. C., Harvey, J. W., Franke, O. L. & Alley, W. M. (1998). Groundwater and surface water—*A single resource: U.S. Geological Survey Circular, 1139*, 79 p., available at http://pubs.usgs.gov/circ/circ1139/.

Winter, T. C. (2007). The vulnerability of wetlands to climate change: a hydrologic landscape perspective: *Journal of the American Water Resources Association*, v. *36*, no. 2, p. 305-311.

INDEX

F

G

H

Q

R

S

T

Y